# CEREALS FOR YOUR HEALTH

A practical guide to cereal grains, explaining their
nutritional and therapeutic value and including a range of
exciting cereal recipes.

# CEREALS
# FOR YOUR HEALTH
## How to Benefit from Organically Grown Wholegrain Cereals

*by*

JEAN-LUC DARRIGOL

*Illustrations by Anne-Marie Dessertine*
*Translated from the French by Marjorie E. Nelson*

THORSONS PUBLISHERS LIMITED
Wellingborough, Northamptonshire

First published in France as *Les Céréales Pour Votre Santé*
© Editions Dangles, St Jean de Braye, 1978

First published in the United Kingdom 1984




© THORSONS PUBLISHERS LIMITED 1984

British Library Cataloguing in Publication Data

Darrigol, Jean-Luc
   Cereals for your health.
   1. Grain    2. Diet
   I. Title   II. Les céréales pour votre santé.
   *English*
   613.2'6    SB189

   ISBN 0-7225-0827-1

Printed and bound in Great Britain

# CONTENTS

# INTRODUCTION

Come with me, forget everything, dress practically... We are going to the extreme north of the Indian province of Kashmir, where Tibet, the Pamir Mountains and Afghanistan meet. It is a mountainous region with many peaks over 23,000 feet (7000 metres): a real mixture of high mountains, deep gorges and dizzy cliffs. An unwelcoming area, a severe climate, in absolute geographical isolation. You must walk for a month alongside the mountains on a narrow and dangerous track, before reaching the mysterious country where I am taking you: a hundred villages built on terraces, at the bottom of a valley which is unlike all the others. A valley where illness is unknown. It is the country of the Hunza.

The numerous scientific observations, both rigorous and objective, carried out in the last fifty years by biologists and western doctors, lead to the same conclusion: the Hunza people do not know illness. They are a 'phenomenon' whose state of health is unbelievable to us. The Americans speak of their 'super-health'. The men procreate well into their nineties, work in the fields until they are 110, and sometimes live much longer... to live to 120 years old is not rare among the Hunzas. Amazingly resistant, the exceptional rigours of their geographic and climatic environment have no effect on them: they do not suffer anxiety, fear, doubt or fatigue. You might say they are capable of superhuman feats: heavily laden, barefooted, they can trek for several days over rocks, cliffs and ravines, practically without rest and just a few dry apricots for food.

There is certainly a reason why the Hunza are never ill,

and all the research has come up with the same conclusion: their diet, practically unique in its severity and simplicity, protects them from the onslaught of germs. For those of us who can see certain diseases becoming more prevalent every day (cancer, diabetes, heart disease, rheumatoid arthritis, ulcers, obesity) and seemingly increasing with the rise in the standard of living and the changes in our diet, this 'trip' to the country of the Hunza is a wonderful lesson in simplicity and history. We have lost the sense of proportion, of equilibrium and of control over our bodies; contemporary man, in his world of overeating can see his health running away like sand between his fingers. But how do the Hunzas stay so amazingly healthy?

The Hunza people are totally self-sufficient. They eat exclusively the products they have grown themselves and do not have anything from outside their valley. For them nothing is imported, refined or preserved. It is a wonderful example of symbiosis between man and the land. In the context of this strict dietary self-sufficiency, the people of Hunza are remarkably moderate and frugal. They are satisfied with very little food much to the astonishment of western observers, given their exceptional physical resistance. And the most contradictory thing seems to be the fast observed by the Hunza each Spring: this fast lasts for several weeks and coincides with the period when the Hunza work hardest (ploughing, sowing, the maintenance of irrigation channels worn away by a long winter). Thus it is just when the Hunza have the most work that they eat the least. This long Spring fast cleans out their bodies and allows all the toxins accumulated throughout the winter to be eliminated. It is a purification process.

The Hunza are practically vegetarian. They very rarely eat meat, because they do not hunt very much and have a deep respect for animals. Their basic diet is made up of cereals and fruits with a little cheese and curdled goat's milk. They eat many fruits, in particular apricots. They plant apricot trees wherever the hillsides are not too steep. The apricot is a fruit which has two advantages for them. They extract oil from the kernels and they can dry the fruits giving them reserves for the whole year.

The Hunza mainly eat cereals; they cultivate wheat, barley, millet and buckwheat. With the millet and buckwheat they make porridge, and with wheat and barley they make griddle cakes. They make bread, biscuits and porridge with the flour (wholemeal, of course) which is milled just before use – these are the basics of their meals. Sometimes they eat wheatgerm (they germinate the wheat in the humid sand).

I have called this book *Cereals For Your Health*. I could not resist, as an introduction to such a subject, taking you to this valley where illness is quite unheard of . . . and where the people essentially live on cereals. Remember this lesson. My aim throughout this book, is to encourage you to eat much less meat, sugar and animal fats . . . and far more fruit, vegetables and cereals.

# 1.

# CEREALS

Cereals are the fruits of various grain-bearing plants:

- Wheat (*Triticum vulgare*)
- Rice (*Oryza sativa*)
- Barley (*Hordeum vulgare*)
- Oats (*Avena sativa*)
- Rye (*Secale cereale*)
- Maize (*Zea mais*)
- Millet (*Millum effusum*)

These are the most commonly known and will be discussed in this book.

A plant which does not belong to the graminaceae but is usually classed with them is Buckwheat (*Polygonum fagopyrum*), and I have included it in this study.

Around 3.8 per cent of farming land is used for cereal cultivation, for an annual world-wide output – obviously varying year by year – of 1500 millions of tons.

Cereal production has practically doubled since the end of the Second World War. The extension of cultivated land has been fairly small so it is the yield which has increased, thanks largely to progress in genetics leading to the discovery of varieties better suited to difficult climatic conditions and types of soil. Each cereal has its own value and fills the dietary needs of the people living in the geographical region where it is grown; rice in South-east Asia, wheat in Western Europe, maize in warm areas, oats in the colder areas of Northern Europe . . . we will see why in Chapter 3.

A slump in cereal consumption can be seen today in

Western countries with a high standard of living. Trade is concentrated between these countries and those where cereals are the staple diet. The United States and Canada each year export vast quantities of cereals, particularly to the USSR, and one could say that the international cereal market plays a part in some of the larger political problems of our age. In the future this trend can only become stronger because of demographic expansion in the countries of the Third World and their inability to satisfy their greatly increasing needs. Some of these countries are involved in wars which have paralysed production, leaving them vulnerable and dependent on foreign aid . . . which is, politically speaking, rarely disinterested.

Only a fraction of the cereal grown each year is destined for human consumption. Industrial demands are increasing but it is especially in the area of food for livestock that the expansion is important. The industrialization of meat production (in particular pig and poultry farming) is the root cause of considerable development in the cultivation of maize and barley. Paradoxically, one could say that western pigs and chickens are better nourished than many people of the Third World.

## Food Value of Cereals

The extraordinary success of cereal cultivation in the history of mankind is explained above all by the ease with which one can cultivate, transport and keep them. We will see: in the course of the book, the dietetic interest of cereals, their glucid and protein contents are very close to the dietary needs of humans; their polyunsaturated lipids guard against the formation of cholesterol; they are rich in mineral salts and trace elements; they contain all the vitamins we need. You can now see why they have an exceptional place in man's diet.

Cereals, thanks to their kernels, are above all a source of glucids and therefore calories. And they are very cheap in relation to their calorific value.

Their role has always been to satisfy the largest part of man's energy needs; a 10oz (300g) portion of bread has

# WHEAT

850 calories, which is a third of the daily adult requirement (2500 calories). If calorific needs increase, for example for a manual worker, one only needs to increase one's bread intake.

In the countries of the Third World, the largest part of cultivated land is given to cereals because they contribute the largest part to each man's daily dietary needs. In Asia and in Africa cereals form more than 70 per cent of the daily alimentary diet against less than 20 per cent in Western Europe!

In western countries our diet is nowadays more varied, and the decrease in cereal consumption is in keeping with the massive increase in sugar consumption. It is clear that the sugar from cereal starch is far better for one's health than the white, refined, unnatural sugar which westerners today – and particularly children – eat all day long. We have forgotten the lesson of history.

### The Place of Cereals in History

Stone Age man, surrounded by a hostile environment, had to eat to survive. He did this by cultivation. Among the foods that nature offered him, which included fruits, grains and roots, were wild cereals. These precious wild grains have a great advantage over those of all the other plants because they keep well and can be stored for future use and they quickly satisfy hunger. The importance of these grains increased forty-fold during the Ice Ages when the flora became more rarified because of the changing climate.

With the transition from simple cultivation to agriculture during the neolithic revolution, about 8000 years ago, grains were quite naturally the first plants sown by man. Then, having conquered fire and mastered the art of pottery, Man carried out four types of food preparation with the cereals he had cultivated; roasted corn, soup, gruel and biscuits.

One day a man forgets a biscuit and when he finds it the next day it has altered in its appearance and swollen up like a balloon. The man decides to bake the fermented biscuit anyway and he has just invented bread!

In the same way beer was discovered when a jug of barley gruel, forgotten in some corner, was transformed by accidental fermentation.

The various civilizations of the Ancient World each gave a particular boost to the cultivation of cereals. All the vestiges of Egyptian civilization that we know bear witness to the importance of cereals. They were at that time the centre of active commerce. The Egyptians had almost the same techniques of breadmaking as we have!

There were more than 300 bakeries in Rome at the beginning of Christian times. The Romans were then thrifty people but when decadence set in their eating habits changed: this was the era of great banquets, notorious orgies, where cereals lost their pride of place to thousands of exotic dishes, each one more choice than the last . . . and more harmful to one's health!

Modern and contemporary history shows us an astonishing thread running through all our revolutions. The principal claim of men in revolt is for more bread. It is a sign that without cereals the evolution of humanity would certainly have been different.

Harvesting and threshing, in every latitude and in every age, are accompanied by a cult of songs, ritual dances and offerings to the Gods. These popular traditions have now almost disappeared from the life of western countries. The giant combine harvesters no longer really have anything poetic about them!

## The Use of Cereals

The products deriving from cereals are used today in two types of industry.

—The industries of primary transformation, or agricultural industries: the milling trade, rice and maize mills to produce semolina, starch, glucose, malt and foodstuffs for livestock.

—The industries of secondary processing or food industries: bakeries producing biscuits and rusks, manufacturers of pâtés and soups, breweries and the making of dietary products.

Within the scope of this book only the food industries

interest us. We will see that the more complex forms of cereal usage (rusks, biscuits) have a tendency to replace the simple forms more and more (bread, flour, etc.). A hundred years ago bread provided more than two-thirds of our daily calorie requirements and provided three-quarters of a family's meals. Now cereals account for less than a third of our calorie requirements and constitute far less than 10 per cent of our daily meals!

# 2.
# BE ON YOUR GUARD

Five deaths, dozens of men attacked by hallucinations and eventual madness . . . In August 1951, in the little town of Pont-Saint-Esprit (Gard) an extraordinary phenomenon of mass poisoning decimated a part of the population. Under pressure from public opinion the investigators quickly drew their conclusions and blamed bread for the outbreak.

It was immediately thought to be ergotism (a disease of rye) which produces the same symptoms as those experienced by the afflicted. (This mass poisoning recalls the 'Mal des Ardents' which caused havoc in the Middle Ages amongst the poor who had eaten bread made with rye which was contaminated with ergot.) The law took charge of the affair, but the enquiry, after three years of laborious work, ended in August 1954 inconclusively. Now, the toxicological experts have, after many years of research, returned their verdict: the finger points to phenylmercury acetate – a product used in the preservation of corn.

This same product, which can cause serious foetal malformations – the infamous disease of Minamata – was responsible more recently for several hundred deaths in Iraq in 1972. It was said that the millers could not have ground the grains properly, it is evident that the error was a human one. In this case, it would have been better for the people who ate the bread if those grains had never been treated. Other examples of poisoning have been blamed on human error – including the case of talcum powder containing overdoses of hexachlorophene!

Five deaths in Pont-Saint-Esprit, perhaps five hundred

in Iraq, but how many thousands in Bangladesh? In 1977, starving people looted a train carrying treated cereal seeds. Little was said about the events far away in Asia but what exactly are the dangers involved in treated seeds?

## Cereals and Pesticides
The substance incriminated in the Pont-Saint-Esprit case is particularly poisonous; two grams are enough to kill an adult. Moreover, it has dangerous side-effects on pregnant women.[1] But from grains of wheat to the finished loaf of bread there are many other dangerous substances used in the cultivation of cereals and subsequent processes.

**The seeds.** About 2000 tons of phenylmercury acetate and methyoxy ethylmercury silicate are used each year to ensure that seeds store well, despite the fact that mercury derivatives are extremely poisonous.

**Soil treatments.** Numerous pesticides are authorized in cereal cultivation and in particular 'lindane' and 'parathion'. For 'lindane' legislation in France has fixed the maximum dose in bread-making flour at one milligram for one kilogram of flour. Yet in East Germany the authorities have fixed the limit at 0.1 milligram for one kilogram – ten times less!

**Weedkillers.** Over forty weedkillers are authorized.

**Insecticides.** Used in the treatment of the parts of cereals above ground.

**Grain storage.** In the hulls of ships, silos, mills, there is now a trend to 'gas' grains to ensure they keep perfectly.

**Flour preservation.** Several insecticides are used today in the preservation of flour. It is interesting to note that the law regarding these chemicals varies from country to country. What are the criteria for these decisions?

In the United States other preservatives are authorized. There is even an ultra-sophisticated pesticide Warfarin, an

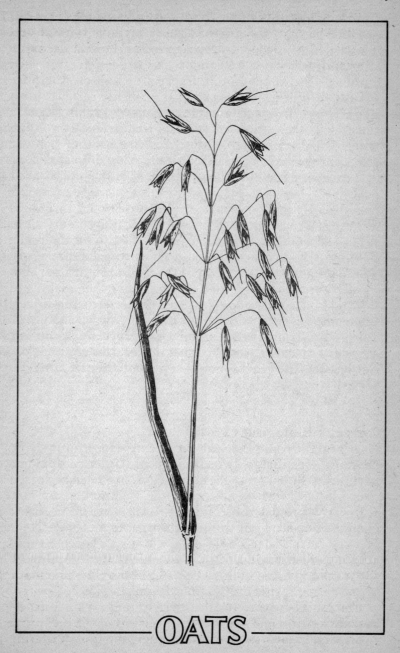

OATS

amazing anticoagulant for rodents. In France they are not satisfied with preservatives for treating flour, they bleach it with chlorine gas, azote trichloride, benzol peroxide and chlorodioxide. And today – as if pesticides are not enough – we are beginning to use a technique still more radical to fight the mites and maggots which attack flour: **Radiation**. An electric current is used which causes a tension in the centre of the flour of several thousand volts, which completely sterilizes it. Gunther Schwab said appropriately of radiated flour, 'This flour is as dead as a sack of cement'. Bon appetit!

**Breadmaking**. The legal definition of bread in France authorizes, in addition to the four basic ingredients, bean flour, ascorbic acid, malt and gluten. Now we are actually helping the invasion of 'special' pre-packed breads. These breads, which represent for the moment 5 per cent of sales, escape the legislation which rules the making of normal bread and of course, contain numerous additives: preservatives, emulsifiers, textural agents . . . and even colourants. Hence we seem to have a very serious problem. The grains used in bread are chemically treated at a risk to our health! And yet this book is entitled 'Cereals For Your Health'.

### Only Eat Biological Cereals

It is perfectly possible to grow cereals without using chemical fertilizers, weedkillers or other pesticides. Several methods of organic cultivation have been perfected over the last few years, each day winning over new converts amongst those farmers aware of the need for change. The study of these agro-organic methods is not our concern, for it would need a book devoted to that alone, we will limit ourselves to succinctly summing up the method perfected between 1920-30 by the precursor of modern organic cultivation: Raoul Lemaire.

His method rests on the enrichment of manure by decomposition and organic fertilization of the soil using the special features of lithothamne calmagol, a marine algae particularly rich in mineral salts and trace-elements.

To enrich the soil with natural nitrogen, Raoul Lemaire recommended the systematic association of lucerne, clover sainfoin in all cereal cultivation. When Raoul Lemaire opened his first bakery in Paris, in 1931, he enjoyed immediate success. It is true that his wholemeal bread, made from organically grown wheat, and made with fresh yeast and baked over wood fires, had an amazing taste.

It is necessary to emphasize that we ought always to choose wholewheat bread made from organically-grown grain, for it is in the peripheral layers of the grain that the pesticides accumulate. It could, therefore, be dangerous to eat wholemeal bread which has not been made with organically-grown wheat, which can be bought from health food shops and certain bakeries. You should only choose organic products made from cereals cultivated without pesticides and weedkillers, and seeds without chemical preservatives, and processed without synthetic additives. Once again you will find these organic products (bread, flours, cereal flakes, rusks, biscuits, etc.) in health food shops.

[1] A. Roig, in his *Dictionnaire des Polluants Alimentaires* indicates that phenylmercury acetate is one of the most dangerous substances associated with birth defects.

# 3.
# THE EIGHT CEREALS

## 1. Wheat

The history of wheat begins with the history of humanity. Grown since neolithic times in Europe, it has always been – and still is – the dietary basis for the people of temperate climes. In fact, it is the cereal best adapted to our own climate, and the one that corresponds best to our needs. The balance of our diet rests on the regular consumption of wheat as it contains:

- All known mineral salts (sodium, calcium, potassium, magnesium, silicon, phosphorus, sulphur and iron...)
- Numerous trace elements (manganese, copper, zinc, iodine ...)
- Indispensable vitamins ($B_1$, $B_2$, $B_{12}$, D, E, PP ...)

We will go into the properties of each of these elements in the chapter on wheatgerm, because germination purifies all of them.

In all cases, in whatever form wheat is consumed (in grains, flour, flakes, in a loaf of bread, rusks, etc.), it must be wholemeal and organically grown. The grain of wheat must always be used wholly, for in separating the different elements which constitute the grain, one commits an act against nature which can affect our health. The study of wheat grains shows the complementary nature of its various parts.

The grain of wheat, as with all grains, is the result of the development of an ovule in the ovary of a flower, which is at the base of the pistil, on the receptacle of the flower. The receptacle is itself carried by the stalk of the flower. The grain has an oval body, with an eye at the tip, and very fine hairs at the other end. It weighs about 40 milligrams and is divided by a lengthwise crack into two lobes, which are roughly equal in size. In this crack a stalk is found which fastens the grain to the ear.

The grain of wheat is made up of four parts:

**The pericarp.** This is the external envelope of the grain comprising three skins (epicarp, mesocarp and endocarp) and which has a fibrous structure made of cellulose. This external part of the grain is generally called the **bran.** A whole chapter is devoted to this (Chapter 7) as it plays such an important part in our diet; an importance which we are rediscovering today, in particular for those suffering from constipation.

**The protein layer.** Found beneath the pericarp, this part which surrounds the kernel is formed from large granular cells which are very rich in protein. These are the aleurone cells.

**The germ.** Often called the embryo, it is from this that the roots and the stem spring when the grain is planted in the earth. The germ is attached to the grain by the scutellum. This is the most nutritious part of the grain. Certain nutritionists have called it appropriately the *miracle food*. We have devoted the whole of Chapter 6 to wheatgerm.

**The kernel.** Still called the endosperm, the kernel constitutes 85 per cent of the wheat grain, as opposed to 12 per cent for the external parts (pericarp and protein layer) and 3 per cent for the germ (and the scutellum). The kernel is formed from grains of starch encased in a web made of particles of protein: gluten. Gluten is especially abundant in the external part of the kernel. The kernel is quantitatively the main constituent of the wheat grain.

The uses and properties of wheat are largely expanded upon in the chapters given over to bread, wheat germ, bran and to the products of processed cereals. Thus we will not go into them here. We will limit ourselves to one example only, that of an excellent food for revitalizing health, re-mineralizing, and for giving to children and people suffering from over-tiredness; a gruel made from coarse wheat. For one person, grind 3 tablespoonsful of wholewheat (it is essential that the wheat be freshly milled) and dilute this with 6 tablespoonsful of water and one teaspoonful of puréed kernel. This should be eaten on an empty stomach and chewed carefully.[1]

## 2. Rice

Rice is the most prevalent cereal in the world and it is the principle food for half of mankind. Its power to satisfy hunger is a considerable advantage. Fifty grams of rice fills you up as much as 200 grams of bread.

Before it is sold and eaten, rice undergoes several processes:

**Husking.** The rice which falls from the panicles is called rice paddy. The husking process passes the rice paddy between grindstones (or rather, nowadays, rollers) to obtain the rice cargo, thus called because it is shipped in this state.

**Polishing.** This operation, achieved by friction between vertical cylinders trimmed with sheepskin, strips grains of the pericarp, the protein layer and the germ. Polished, white rice consists of nothing but the kernel. In these two processes (husking and polishing) the rice loses more than 30 per cent of its initial weight, 80 per cent of its lipids, 60 per cent of its mineral salts and practically all its vitamins. This treatment of rice, this refining process, is critical to its nutritional value: vitamin $B_1$ being eliminated, consumption of white rice brings with it a serious vitamin deficiency in those populations who rely on rice as their main food.

The study of the disease beriberi is at the heart of the discovery of the role of vitamins in human nutrition.

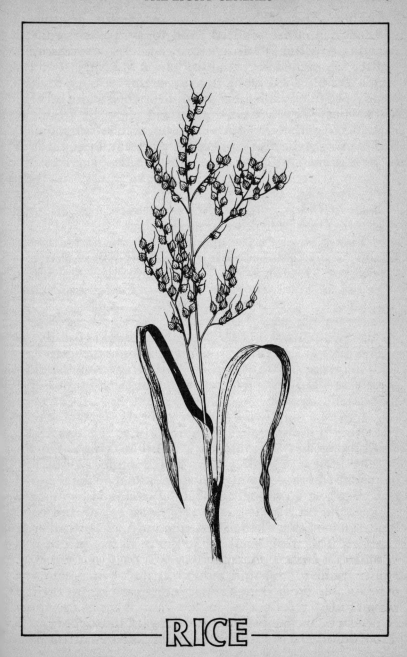

RICE

Beriberi caused havoc in Asia before the benefits of vitamin $B_1$ were realized. And today, paradoxically, a return to eating whole rice (brown rice) is not advocated. Rice is enriched in synthetic vitamin $B_1$. Thus we have a husked, polished, white rice (the grains are coated with a mixture of talc and glucose) and enriched with synthetic vitamins. We really do live in a sad age!

It is clear that those people desirous of reforming their diet must begin, in every case, by systematically eliminating basic refined products (cereals, oils, salt, sugar, etc.) and by replacing them with natural products, whole cereals, oils, sea salt and brown sugar. As regards rice in particular, one can find organically grown brown rice in health food shops. You should always choose this.

Rice is low in fat compared with wheat which must make it preferable to other cereals for invalids. It has hypotensive properties, and its relative lack in sodium and potassium make it an ideal food for those suffering from oedemas or from heart or renal complaints. Thus it is recommended – at the exclusion of all other food – in the Kempner diet; in cases of high blood-pressure, nephritis, in certain heart conditions or in the last stages of pregnancy, in the case of oedemia. The Kempner diet consists of eating only rice (300 grams a day) with a little fruit juice.

Less rich in mineral salts than wheat, rice can be used more without risk of getting 'bound-up'. It is chosen for slimming diets in preference to other cereals. On the other hand, its lack of iron is the principal cause of anaemia in those countries where it is the staple diet.

Rice is often wrongly thought of as a food which causes constipation. This is quite wrong because only the water used in cooking the rice is astringent. You can drink rice water, made from cooking 20 grams of rice in a litre of water, in cases of diarrhoea.

To add to the information given on rice, Ohsawa advocates crude rice against intestinal parasites: eat a handful of crude rice in the morning on an empty stomach, chewing each mouthful very thoroughly.

Poultices of rice flour relieve skin inflammations. This

same rice flour, used to make a mask, gives excellent results to blotchy complexions. Equally, 'rice powder' has always been valuable in make-up.

## 3. Barley

Barley is the oldest cereal known: it existed more than 4000 years ago in Abyssinia and Nepal. Its vegetative cycle is particularly short: 60 days from sewing to harvesting, which explains its important place amongst the cereals.

In dietetics one uses hulled barley in preference to pearl barley. The hulled barley is the fruit of the simple hulled grain: the external envelope is removed in a stone mill. On the other hand, pearl barley is refined, polished and whitened like glazed rice. Barley is an excellent reviving food. A barley infusion of 40 grams of hulled barley boiled in a litre of water makes a really thirst-quenching drink.

Barley is used principally in making malt and beer. The malting is a process of controlled germination, followed by a torrefaction of the germinated grains. This germination provokes the appearance of an enzyme which aids the digestion of starches by transforming these into a more simple sugar, maltose. Those who have difficulties in digesting flours should eat malted flours.

Malt is also advocated by dieticians in place of coffee. It is sold in health food shops in grains, powder or instant forms.

Digestive or urinary problems may be soothed by adding barley to soups: the emollient action is most beneficial. Rich in calcium, potassium and phosphorus, barley has a recalcifying effect during growth, periods of mental debility and in cases of mineral deficiency.

## 4. Oats

Contrary to other cereals, oat grains are not arrayed in ears but scattered in tufts. Oats can also tolerate rain at the end of its growth cycle which allows it to flourish in northern latitudes, like Scandinavia. Traditionally, oats have always been a nourishing, energy-giving food providing strength and vigour: the Huns dined on bowls of oat gruel, while the Scots get their legendary physical strength from porridge: not forgetting horses who have always appreciated oats!

Oats stimulate the thyroid gland, and this improves resistance to colds: it is the cereal to eat in winter rather than barley, especially as oats are very rich in lipids. The easily assimilated fats of oats ensure resistance to low temperatures.

Oats contain a stimulating substance avenose, which improves the performance of all those who need a lot of energy; they are recommended to sportsmen during competitions, to manual workers and to lovers! In effect, oats contain a hormone which acts on sterility and impotence.

Another interesting property of oats – hypoglycaemiant – means that they are recommended for diabetics. Those who find it difficult to pass urine might benefit from drinking the product of 20 grams of oats boiled for 30 minutes in a litre of water.

*Note:* Oats also contain purines and eating them may invoke an attack of gout!

A smattering of d'Avena Sativa (40 drops when going to bed) helps insomnia. Legend recommends that nervous people should sleep on a mattress stuffed with oats. Finally, Kneipp gives us a remedy for corns, blisters and ingrowing toenails: a footbath for 30 minutes in water at 25°C/77°F in which one has boiled oat straw.

## 5. Rye

Rye has for a long time had a bad reputation because of a fungus, claviceps purpurea, more often called ergot, which sometimes attacks in humid weather. Ergot contains a very poisonous alkaloid called ergotoxin, which causes an illness called ergotism; this was very widespread in the Middle Ages.

In medieval times it was called fire pest or St Anthony's Pest. It attacks the poor, those who buy contaminated grains because they are sold cheaply. Ergotism attacks the central nervous system: those suffering from it become 'mad'. Luckily we know how to prevent rye ergot these days by using natural methods. So rye flours sold today are always perfectly healthy. It would be mistaken to attribute the widespread poisoning in Pont-Saint-Esprit to tainted rye flour (see Chapter 2).

Rye is a cereal which can be used for breadmaking, indeed it is its main use. The countries of North East Europe eat a lot of rye bread, a trend which is gradually spreading to Britain, yet rye is less rich in gluten than wheat as well as being heavier and less aerated. However, 'rye bread' is usually made from a mixture of flours, most commonly rye and wheat.

It is intriguing that cardiovascular problems are rare in countries where rye bread is mostly consumed. This is because rye is excellent at thinning the blood, softening the arteries and helping in all cases of high blood-pressure, arteriosclerosis and heart problems in general.

On the debit side it must be said that rye contains relatively little vitamin $B_3$ (PP) which is why the vitamin deficiency syndrome known as pellagra has been widespread in the past. Our diets being rich in $B_3$ from other sources, we can quite happily eat rye bread, as a change from our usual wheat bread, especially if we have circulatory problems, or in an effort to prevent them.

## 6. Maize

This cereal, which grows to nearly four metres high, was introduced in Western Europe by the Spanish in the sixteenth century, after the discovery of America. Maize was the basic diet of the Pre-Columbian civilizations. The first varieties of maize cultivated in Europe responded to warmth and humidity which explains why it prospered in Southern Europe. But today, after exhaustive genetic research, we have perfected hybrid varieties which can adapt themselves to different conditions.

Maize cultivation has developed very quickly due to its high yield and the numerous uses maize has in contemporary food industry. For instance, starch extracted from maize is used in making sauces, soups and sweets.

Maize is the cereal richest in lipids and its fat content explains the recent development in the manufacture and consumption of maize germ oil. Maize germ contains 35 per cent lipids and the percentage of polyunsaturated fatty acids in the oil extracted from the germ is very high. 80 per cent of the fatty acids from maize germ oil (in particular oleic acid and linoleic acid) are polyunsaturated, which recommends the use of this oil for its athermanogenic properties, in cases of hyper-cholesterolism. With only 20 per cent of saturated fatty acids (palmitic acid and learic acid), maize germ oil is strongly recommended for cases of excess blood cholesterol. But it is important to note that maize germ oil is not a cold pressed oil. Its extraction is excessively refined, heated and deodourized. It is even purified with chemical solvents used for dry cleaning clothes. The dietetic properties of this oil are thus largely affected by this excessive refining process. Unlike oats, maize is the cereal best suited for warm climates. Because of its moderating action on the thyroid gland one should choose maize in the summer. But one should be careful, for maize, like rye, is virtually devoid of vitamin $B_3$ (PP), so maize can only be a complementary food.

## 7. Buckwheat

This is not a true cereal, nor does it belong to the graminaceous family like the others, but it is generally regarded as a cereal. It has a short vegetative cycle and flourishes in poor, sandy soils.

It contains more calcium than wheat, an important quality for children and mothers who are breast-feeding.

Buckwheat contains numerous essential amino acids amongst which is trytophane, typical of animal proteins. It is therefore a useful complementary food.

# 4.

# CEREALS AND DIETS

## 1. The Hygienist Diet

'The suppression of all cereals is imperative'
A. Mosseri.

In 1928 Shelton opened his 'health school' and began his fight against 'the growing number of doctors and medications which are becoming a grave danger to human life'. This is the advent of the naturist health movement formed because of the 'abuse and incompetence of medicine'.

The hygienists tell us to become masters of our feelings and emotions, to cultivate patience, calm, composure and level-headedness. They place value on the necessity to exercise regularly and sleep regularly. They extol above all the necessity and the virtues of fasting as a means of *detoxification, purification* and of *regeneration*. Finally, they recommend us to only eat when we are hungry, and then only wholefoods. There is no cooking in a hygienist diet, and overeating is unimaginable. Hygienists eliminate all those foods 'contrary to nature', such as meat, cooked meats, sugar, fats, jams and cereals. The excess of protides is the worst excess of all for the hygienist, because cereal wastes are acidic, causing blood-poisoning – according to them – which overworks both the liver and the kidneys.

Their diet rests on the fundamental principle of the need to disassociate foods in order to enjoy harmonious food combinations. One must avoid mixing certain foods which require different enzymes during digestion. Wrong combinations of food, by mixing foods which are not digested compatibly, bring digestive troubles, wind, flatulence, acidity, insomnia, constipation. It is therefore imperative to combine homogenous foods during a meal, as far as the digestive enzymes they require are concerned. So, for example, one must not eat a cheese sandwich – or, for that matter, a ham or any other cooked meat sandwich – because the bread prevents proper digestion of the cheese and vice versa. Supposedly the digestion of the protein in cheese opposes the digestion of the bread.

Not only do hygienists not hold with sandwiches, they reject bread altogether! For them the ideal foods are those of the anthropoid monkey: fruits, nuts (hazelnuts, almonds, etc.) and green shoots. They say the characteristics (anatomical and physiological) of the human digestive tract are adapted to the consumption of fruits; the number and structure of the teeth, length and nature of the digestive tube, position of the nails, properties of the saliva . . . Thus hygienists go much further than vegetarians and vegans; they also omit cereals from their diet, because they accuse them of being indigestible, acidifying, decalcifying; the cause of all ills. Mosseri wrote: 'Man has neither the teeth, nor the salivary glands, nor the stomach, nor the liver, nor the intestine to eat cereals'. The hygienists are not grain eaters, for they do not have the gizzard for crushing grains, which birds possess. Hygienists are extremely harsh about the macrobiotic diet which they accuse of being poisonous and acidifying, responsible for all digestive problems, colitis, diabetes, hepatitis, and also gripe and rheumatism. The only concession that hygienists make to cereals is the consumption of the germ. It is a concession infinitely too modest, and we are convinced that hygienists are wrong.

The hygienist principle which states that man is not a granivore does not stand up to the most elementary analysis: it is obvious birds have a gizzard to crush grains because they have no teeth. Man's teeth, and the chewing process negate the need for a gizzard, which would be superfluous. Moreover, the transformation of starch into more simple sugars which the body can assimilate is made possible by the presence of ptyaline, a digestive enzyme in saliva: thus the hygienist's argument that the human digestive tract is not suited to consume cereals cannot be taken seriously.

What is even more important is that the hygienist diet is particularly deficient in proteins and the B group of vitamins.

Likening man to an anthropoid monkey is no longer valid – if it ever was. Going way back in time man has always eaten cereals and today we can see that the healthiest people, such as the Hunza and Caucasians, eat mainly cereals, a source of strength and longevity. But we must be wary of falling into the trap of only eating cereals – which some people advocate.

## 2. The Macrobiotic Diet

'All illness can be completely cured within ten days by eating only cereals.'

G. Ohsawa

George Ohsawa, who was given up for dead by orthodox doctors, cured himself of tuberculosis and went on to lead a very active life, establishing the basics of macrobiotics, an invigorating and nourishing diet which now has many followers.

Since the turn of the century, our diet has totally changed, with an increase in the consumption of meat, sugar and fats, our traditional eating habits have been overturned by excessive food processing and chemical additives (colourants, preservatives, emulsifiers, etc.). The twentieth century has stimulated all kinds of degenerative diseases such as cancer, diabetes, cardiovascular complaints, and medicine seems powerless to resist. The search for alternatives explains the world-wide success of macrobiotics.

Macrobiotics imposes a radical new life-style and diet in which an oriental philosophy of *wisdom* plays a central role. Followers of macrobiotics must find a new way of living from within themselves. They must live and act in a frugal way, in harmony with nature, in peace, joy and happiness achieved through an improved spiritual and physical health. Macrobiotics leads to the mastery of one's soul, to great changes, to rebirth and to the resurrection of oneself. According to Ohsawa, the 'seven conditions of health and happiness' are:

1. no fatigue
2. healthy appetite
3. deep sleep
4. good memory
5. good humour
6. speed of judgement and execution
7. justice

For Ohsawa everything is possible by having faith in oneself and in one's capacity to be reborn. He wrote: 'Man must be his own doctor'. It is the philosophy of Tao, which in essence is self-improvement leading to self-healing.

Taoism is based on the fundamental dichotomy of the universe, which divides everything into two opposed but complementary forces: Yin and Yang. Everything contains an opposite which produces an equilibrium: day and night, man and woman, war and peace. And so on the dietary plane there are Yin foods and Yang foods. Everyone must try and balance these two types of food. Macrobiotic dietary principles are straightforward: you must not eat foods which have been refined, conserved or coloured, but only natural products. You must avoid refined sugars found in manufactured drinks, sweets, chemical ice cream, etc. and animal fats. You have to reduce your meat consumption and increase your vegetable intake, as well as fruit and cereals.

Cereals are very important in the macrobiotic diet. They represent 50 per cent of food eaten, this proportion increases to 100 per cent in cases of illness, which we shall see later. The macrobiotic diet advocates specific foods, in particular gomasio (a mixture of toasted sesame seeds and sea salt) and tamari (a sauce made from soya). But in each case, cereals must represent at least half the volume of each meal.

The cereals, of course, must be whole and grown

MAIZE

organically. As far as possible, they must be eaten as grains (in preference to flour or flakes) and chewed carefully. The importance of chewing is clear when, in this diet, one is eating raw cereals. A mouthful of raw rice must be chewed 150 times before being swallowed. When using flour one should grind the grains at the last moment to avoid oxidation, which can decrease their nutritional value.

In particularly serious illnesses, as the quotation indicates at the beginning of the passage, it is necessary to avoid all foods other than cereals for 10 days. It is Ohsawa's seventh diet which should be followed, and after an improvement in health, the patient should switch to Ohsawa's sixth diet (90 per cent cereal and 10 per cent vegetables) and so on.

The fundamental position of cereals in this diet can bring about balanced health, but we cannot subscribe to the idea that cereals are a panacea, capable of curing serious illnesses.

## 3. The Vegan Diet

Unlike vegetarians, who exclude animal flesh (except fish) from their diet but not animal derived products (milk, cheese, eggs, honey, etc.), vegans totally reject all animal products.

We will see at the end of this chapter how important cereals are in the vegan diet, but first we should briefly mention the arguments for abstaining from meat.

— Meat has quite a low calorific value: about 220 calories per 100 grams of meat, compared to 360 calories (over 50 per cent more) per 100 grams of wholemeal flour.
— Meat contains very few mineral salts (less than 1 per cent).
— Meat is rich in saturated fatty acids, generating cholesterol and arteriosclerosis.
— Meat produces little intestinal waste: it is a constipating and acidifying food.
— The human body is not adapted to the consumption of

meat, for our liver and kidneys are incapable of filtering (via the liver) and eliminating (via the kidneys) the products of digesting amino acids from animal protein: the liver and kidneys of a man who regularly eats meat are overworked and all sorts of problems arise from their abnormal use.

— The breakdown of meat during digestion leads to fermentations in the intestine and the body is invaded, as a result of the permeability of intestinal mucus, by poisonous waste matter.

— The nucleus of the amino acid cells derived from the breakdown of animal proteins, is rich in purines. The metabolic breakdown of these purines produces uric acid which is extremely harmful, because it invades the body tissues and joints. Uric acid from meat causes rheumatism, arthritis, digestive ills and nervous disorders, uraemia and nephritis. Many of the ailments characteristic of our time are a direct result of the overconsumption of meat.

If we have convinced you of the harmful effect meat has on your health you should also note that meat, eaten moderately, seems to be indispensable for its share of protein; at least that is what the nutritionists think. In place of meat, vegetarians advocate the consumption of protein derived from animals, such as milk, cheese and eggs. But vegans uncompromisingly reject all animal proteins without exception.

If you ask where vegans get their protein the reply is - cereals, of course! Cereals form the basis of the vegan diet. Without them their diet would be seriously lacking in proteins.

Vegans obtain these vital proteins in three ways:
— The first method is to eat large quantities of vegetables. But such large amounts overload our digestive tract which is not suited to such work. For if we are not carnivorous neither are we ruminants.
— The second consists of eating a large amount of starchy foods (dried kidney beans, lentils, broad beans, split peas, etc.) but such an excess would be harmful because it

would lead to many digestive problems: flatulence, dyspepsia, stomach distention.

— The third is the most reasonable: this consists of finding the necessary daily ration of protein in the regular consumption of whole cereals. The protein content of cereals is far from negligible, as we have seen elsewhere. 300 grams of wholemeal bread corresponds – as far as protein is concerned – to 125 grams of meat. For vegans there is no other possible alternative. The protein necessary for good health must be found in cereals.

# 5.
# BREAD

Nowadays it can be confusing when buying bread. We have at least three types to choose from:
— In most bread shops you find white bread with enormous air holes, bizarre cavities and a beautiful golden crust in all sorts of attractive shapes and sizes. But it's a very different story when it reaches the table: pale and insipid, it hardly keeps from one day to the next before becoming rock-hard, a bread you soon have to throw away in disgust.
— In large supermarkets (contemporary temples of food!) beside the freshly baked bread, you can find shelves covered with loaves from modern bread factories, bread possessing the same 'qualities' as the bread described above, wrapped in cellophane. Breads rich in emulsifiers and preservatives, at the expense of taste.
— In certain bread shops, unfortunately still few and far between, and in all health food shops, you can find wholegrain breads, baked with fresh yeast in a wood-fired oven, bread with a subtle, indefinable taste . . . a taste which brings back memories and makes you ask: how could you have eaten, for so long, this dreadful white stuff, so totally lacking in flavour and how can you make sure you can always get wholegrain bread? An impulse which will probably lead you to make it yourself.

Bread has not always existed. Primitive man ate cereal gruels and coarse biscuits. The discovery of bread was no doubt accidental; a forgotten wheatflour biscuit, cooked two or three days after it was mislaid, became distorted and swollen. Man invented bread through the complex and mysterious process of fermentation.

During forty centuries, breadmaking techniques changed only slightly. Then in the twentieth century everything turned upside down. In the name of efficiency, productivity, practicality ... contemporary man chose to make a clean sweep, to do away with his ancestor's methods.

Today, after decades of accumulated mistakes, certain conscientious bakers have reinstated our great-grandfather's methods of breadmaking. Techniques, which two or three generations of neglect did not eradicate, have been revived.

Bread is not just a symbol in man's history. It always represents the basic diet of millions across civilizations, continents and eras. Without going into details we know the Egyptians could make perfect bread: numerous archaeological traces prove it. The Greeks were such proficient breadmakers that they produced seventy-two different varieties! In fact, at the height of the Roman Empire, they had the monopoly on this art: the 300 to 350 bakeries in Rome under the reign of Augustus were all owned by Greek emigrants.

The Romans, through their imperial ambitions, invaded regions which enriched their knowledge in many ways: in breadmaking they learned from the Celts and the Iberians to use the froth off beer (barm) to lighten their bread.

In the Middle Ages the consumption of wheat bread declined in the wake of the spreading popularity of rye, buckwheat or a mixture of wheat and rye breads. One custom was particularly common in this period. Stale bread, cut into large, thick slices, was served as an accompaniment to other foods. Only the crust, soaked in juice, was eaten at the end of a meal.

Wheat bread was not generally found in France until the end of the eighteenth century. The use of barm, which appeared at the end of the seventeenth century, despite the criticisms of the doctors of the day who said it was harmful, allowed a special bread to be made called 'Queen's bread'. It was a long time before the use of barm was abandoned in favour of products from the fermentation of cereals, and then at the beginning of the century, by the yeasts from fermented molasses. Bread consumption has been in decline since the beginning of the twentieth century, as the following table shows.

| | | |
|---|---|---|
| 1900 | : | 700 grams per person per day |
| 1920 | : | 500 ,, ,, ,, ,, ,, |
| 1938 | : | 400 ,, ,, ,, ,, ,, |
| 1950 | : | 250 ,, ,, ,, ,, ,, |
| 1977 | : | 160 ,, ,, ,, ,, ,, |

A drop in the 'quality' is not the cause. It is not the fundamental reason behind this trend, but if the bread were better, perhaps those who hardly eat it any more would voluntarily find their way back to the bakers!

The rising standard of living, a diversified diet and a more sedentary lifestyle can be used to explain the decline in the importance of bread in our diet. Bakeries can be found nearly everywhere but their closure in small villages symbolizes the distress of the rural world affected by the exodus of young people. Happily the reverse is now happening. One now finds, at exhibitions and fairs, bread ovens designed for domestic use and ecological publications giving advice on making your own oven.

## How to Make a Good Loaf
The legal definition of bread is as follows:
Today's bread is the product of the baking of a dough composed exclusively of breadmaking flours, salt, yeast and water. We are able to add to the legal definition that the dough destined for baking has been shaped after kneading and fermentation.

The basic recipe which reigns supreme in traditional French breadmaking is unchanging:

100 parts flour
60 parts water
2 parts salt
1 part yeast

Besides this, French law allows the use of four additives:

broad bean flour (less than 2 per cent)

ascorbic acid (vitamin C)
malt (less than 0.3 per cent)
gluten

We have seen that the making of special pre-wrapped breads seems to escape this regulation since other additives are used. One is the omnipresent preservative in this type of bread – calcium proprionate. The authorized additives each have their own distinct properties. The broad bean flour whitens the flour. This is the only means of whitening bread allowed in France. Vitamin C helps the fermentation, as does the malt. The malt also helps to colour the crust. Finally, gluten is sometimes added when making special breads containing less glucids for diabetics.

In order to be fair, we must underline that the raising agent used in breadmaking is not a chemical one, as it is often wrongly thought. It is baker's yeast, which is quite simply a perfectly natural, microscopic unicellular fungus – *Saccharomyces cerevisae*. However, pastry makers do use a chemical raising agent, which explains the confusion. Leaven is obtained by keeping back some fermented dough and incorporating it in new dough.

It is worth noting that some bakers use both yeast and leaven mixed in varying proportions to make a 'mixed' bread. Leavening bread takes 16 hours, the use of baker's yeast considerably shortens the night's work and allows the baker to keep regular hours, which is not possible when working with leaven, which requires irregular fermentation periods according to temperature and humidity.

To complete this definition of bread, it should be said that two types of ovens are used. The direct heat oven and the indirect heat oven. In the latter the cooking area is separate from the source of heat, whether it is wood, electricity, oil or gas, the result is the same. In a direct heat oven the sort of fuel used is very important in that it is in direct contact with the bread. In a wood-burning direct heat oven the embers and cinders left by the burnt wood are removed before the bread is put into the oven. The bread then bakes from the heat stored in the oven walls.

At the beginning of this chapter, I emphasized the mediocrity of the bread we are offered today. I shall, of course, show how we can rectify this sad state of affairs. But before doing so I must describe in what conditions bread is made in other parts of the world – take for example the United States; the French rightly complain that their bread is getting worse and worse, and yet, ironically, it is appreciated by visitors to France.

The legal definition of bread in the United States bears no resemblance to the French version. American bakers, or should we say chemists, use these ingredients in the following proportions:

100kg of flour bleached by a 15000 volts electric current or by adding chlorine, benzene peroxide and chlorine dioxide

65 litres of water

2 kg of yeast derived from aluminium

2 kg of salt

1 kg of malt

6 kg of sugar

4 kg of fat

5 kg of powdered milk

0,500 kg of 'yeast food', a mixture of potassium bromate, ammonium chloride, potassium tartrate and amylase HCl

— monostearate of polyoxyethane in indefinite quantity, so the bread stays fresher longer

— diglycol which softens the texture of the bread.

This list is somewhat frightening and should make us wary of pre-wrapped bread carrying American slogans! It is obvious that American bread has absolutely nothing to do with the natural product which should still be thriving in the U.S.A. Therefore it is hardly surprising that so many Americans have digestive troubles, allergies, obesity, eczema and hair loss!

This example shows us how dangerous the revolution in breadmaking has been and must convince us of the grave necessity of opposing modern industrial techniques which are replacing traditional methods.

The baker must remain a craftsman. We must safeguard this ancient craft from the tentacles of industry. Moreover, we must incite bakers to rebel against complacency in the face of this onslaught. We should push them into returning to processes which will produce a baking renaissance – recreating the marvellous flavours of time gone by! To make a good loaf, five conditions are necessary:

1. The wheat must be organically cultivated without chemical fertilizers and without weedkillers, from seeds which have not been treated with preservatives, and the harvested wheat must be transported and stored without coming into contact with chemical preservatives.

2. The flour from this wheat must be made in mills equipped with stone wheels, giving a slow milling without heat, without using rollers. This flour must also be stored without preservatives.

3. The flour used to make the bread must be absolutely whole and not a white flour.

4. The breadmaking process must use leaven and not yeast.

5. The baking must be done in an oven with direct wood heat.

A few readers will smile at these lines. Some farmers are incredulous at the thought of being able to produce organically, sufficient harvests and effectively fighting off the attacks of parasites or cryptograms. These farmers are wrong because organic farming is a marvellous alternative which takes into consideration the ecological needs of our environment and the future of generations to come. Professional millers will accuse us of being old-fashioned and will oppose the return of stone mill wheels on the grounds of efficiency. These manufacturers are mistaken, for the dietetic quality of flours produced by their rollers is unquestionably mediocre. A return to the techniques of the past would do them credit. Some bakers who wish to share in our leisured society are reluctant to re-learn the methods of the past. These bakers are short-sighted because in years to come there will be an increasing demand for wholemeal bread made with leaven and baked with wood heat. A demand which will be so great that a day will come – and not too far away – when wholemeal bread will be more in demand than white bread. The change is underway and gradually farmers are returning to organic agriculture, the conscientious manufacturers are returning to the days when they were still millers, and bakers were true craftsmen.

Some people will challenge our preference for wholemeal bread. They think wholemeal bread destroys calcium, a mistaken belief prevalent several years ago. They are wrong for the following reasons.

## The Necessity for Wholemeal Bread

Certain people accuse wholemeal bread of causing anaemia and calcium deficiency in those who eat large amounts of wholemeal bread and who never eat white bread. They base their argument on the presence of phytic acid in the peripheral coats of the wheat grain, layers which are mixed with the kernel in the milling process to produce wholemeal flour. This phytic acid present in wholemeal flour is a combination of inositol and phosphoric acid. It has been proved experimentally that phytic acid interferes with the metabolism of calcium and iron. Phytic acid combines with the calcium to form an *insoluble* precipitate, calcium phthalate, which is eliminated with other body wastes: the assimilation of calcium is thus impossible in the presence of phytic acid. And regular and exclusive consumption of wholemeal bread leads to a serious calcium deficiency. Also, phytic acid hampers the assimilation of iron by forming ferrous phytate which is equally insoluble. These are serious arguments if one sticks to the strict chemistry involved and the laboratory experiments, carried out on isolated elements out of their natural context. It is here that the error is to be found: it is no use at all that chemists reproduce elementary reactions between phytic acid and calcium *out of their natural and living environment* where they continually interact: i.e., during breadmaking. In reality – we can never stress the necessity of appreciating the totally original character of the natural product enough – it is certain that flour contains a catalyst which *neutralizes the action of phytic acid during fermentation of dough: phytase*.

The presence of phytase entirely changes the problem: phytase hydrolizes the phytic acid by separating the molecules of inositol from the molecules of phosphoric acid, and the action of the phytic acid on the calcium and iron is thus counteracted and reduced to zero. Nature has organized things once again and secret reactions which happen inside the dough whilst it is fermenting have absolutely nothing to do with those which the chemists are able to reconstruct in their labs. But we must underline one essential point, which can prove to some extent at least, that the detractors of wholemeal bread are right: to take effect, phytase needs time, because the hydrolysis of phytic acid is not instantaneous but happens gradually, over several hours. With traditional breadmaking methods using leaven, fermentation lasts a whole day. The phytase has plenty of time to act and the phytic acid is neutralized, which safeguards the calcium and iron. This observation is a point in favour of *slow fermentation using leaven*. But when making bread with baker's yeast, fermentation is reduced to half an hour in cases where dough is kneaded very swiftly! Not only does the phytase not have time to act, the phytic acid is not neutralized and combines, as we have described, with calcium and iron to give an insoluble precipitate. Rapid fermentation thus absolutely forbids the use of yeast when making wholemeal bread since to do so would run the very serious risk of causing calcium deficiency in those who eat it.

On the whole, one can sum up this paragraph by saying 'The making of wholemeal bread necessitates a slow fermentation using leaven'.

Where leaven is used, far from causing calcium deficiency, wholemeal bread contributes to keeping us in good health because of its richness in mineral salts – as opposed to white bread – and vitamins, which help break them down so the body can assimilate them.

'Wholemeal bread made with leaven is a powerful remineralizing agent.' Be careful when buying your wholemeal bread; always ask whether:
—it has been made with a wholemeal flour made from organically grown wheat.
—it has been made with a slow leaven fermentation.
These two points exclude wholemeal bread made with yeast and normal, not organic flour. They equally exclude bread which is called 'wholemeal' through the addition of bran to white flour: this type of bread dupes the public although it is easily recognizable because of its texture.

We explore the properties and virtues of cellulose, in which wholemeal bread is particularly rich, in Chapter 7.

## Methods of Breadmaking
In 1962 the National Centre of Scientific Research (NCSR) published an important work of over 1000 pages on the problem of maintaining the quality of bread, under the direction of Professor Terroine. His conclusions run contrary to the criticisms made by the nostalgios of an age when breadmaking methods were more like a craft and the bread itself more natural. An explanation of these breadmaking techniques will give us a better understanding of this evolution.

Dough for breadmaking is made, as we have seen, from flour, water, salt, yeast (or leaven) in the following proportions: 100-60-2-1. A clear understanding of the chemical processes which alter the structure of the flour during breadmaking calls for a more exact knowledge of its constituting elements: flour is made up of glucids (principally starch), proteins (gluten in particular), lipids and enzymes favouring certain metabolisms.
—**Starch** is the most important element, quantitatively: 50 per cent of whole flour, 75 per cent of white flour. Its enzymatic breakdown calls for the intervention of the amylase present in saliva, called ptyaline, and of pancreatic amylase. It is thus dextrinized, broken up into more simple sugars, of which maltose is one.
—**Gluten** is composed of two amino acids: gliadin and glutenin. One should specify that the peripheral pockets

of the wheat grain, and therefore wheatmeal flour as well, contain an enzyme which reacts with gluten causing it to lose some of its elasticity: glutenase. Thus the gluten's retention of gas bubbles which occur during fermentation, is counteracted by the effect of the glutenase, and we can see why wholemeal bread is less porous, less aerated and generally more substantial than white bread.

—**The lipids** in the flour are modified in the course of fermentation by the action of lipase and lipoxydase which liberate fatty acids. These are responsible for acidifying the dough.

During breadmaking, the complex upheavals affect the dough throughout the different processes which lead to the finished loaf: kneading, initial fermentation, shaping into loaves, second fermentation, baking and cooling.

## (a) Kneading

The kneading process carefully mixes the various ingredients so the dough becomes perfectly homogenous. In times gone by, the kneading was done by hand. Today bakers use mechanical kneading troughs, kneading the dough at ever-increasing speeds (90 turns per minute for 20 minutes). With these modern machines, the bread is lighter and more voluminous, but the taste suffers. During kneading the water is thoroughly mixed with the gluten and starch. The quantity of water required to make a dough of pliable consistency varies according to the type of flour used. In fact, breadmaking flours have a differing capacity to become hydrated. This is what is called the *strength* of the flour. Strong flours make much better bread. All the work put in to improve wheat varieties, has perfected wheats which give strong flour. Don't add too much water and don't knead the dough for too long or it will become a solid and sticky mess. It is gluten which is responsible for the elasticity and pliability of dough. This is the exceptional property of wheat gluten which accounts for the excellent baking quality of its flour. The majority of cereals are unsuitable for breadmaking because they do not contain a gluten with the right properties to encourage fermentation in the dough. Thus, to make rye bread, it is necessary to use an equal amount of wheat flour and rye flour.

The air incorporated in the dough during kneading plays an essential part in promoting certain indispensable oxydations in the dough mixture. The use of ascorbic acid is legally authorized, as we have seen, to improve the oxydation processes. Ascorbic acid, paradoxically, has reducing properties. But its incorporation in the dough causes enzymatic changes which transform it into an oxidizing agent. Moreover, the addition of broad bean flour, authorized for whitening flour and increasing its strength, improves the oxidation processes, for the bean flour contains some lipoxygenase.

## (b) Fermentation

When the dough is well-mixed after kneading, leave it to rest. The fermentation then begins, a complex enzymatic process of biological and physical chemistry which varies in duration according to the manufacturing process; if you use leaven, the fermentation takes a whole day, with traditional yeast it takes four hours. Modern methods of kneading have reduced fermentation to a mere half hour.

Schematically the fermentation produces carbon gas by transforming the starch sugars; the bubbles of carbon gas are retained by the dough because the elasticity of the gluten allows the dough to inflate but not burst. From the start of fermentation, there is a simultaneous reaction between the amylase of the flour and the enzymes in the yeast. This is essential because the enzymes separate the products of the broken down starch. The amylase attacks the starch, transforming it into maltose. As the addition of malt is authorized by law, this attack on the starch is made easier. During this stage, the yeast is causing a fermentation of an alcoholic kind in the dough, which produces two new essential elements: carbon gas and alcohol. This alcoholic fermentation depends on simple sugars which appear gradually as the starch is broken down and the chemical reactions take place. These sugars are maltose, glucose, fructose and saccharose.

We must emphasize here that at the time of the initial fermentation, the formation of alcohol is predominant in relation to the carbon gas. But after the loaves are moulded and up to the moment of baking, during the final fermentation, it is carbon gas which is produced in the greater quantity.

The temperature at which alcoholic fermentation happens is very important for the entire process. The best average temperature for a normal fermentation is 25°C/ 77°F. By raising the temperature to 35°C/95°F, the fermentation is twice as fast. On the other hand, if one cools the dough to roundabout 10°C/50°F, the fermentation is impaired considerably. Bakers nowadays tend to cool the dough. They stop fermentation after the first stage, by shaping the loaves and cooling them to 5°C/40°F. This technique allows them to programme their work time, for they can prepare dough for the next day as they need it.

The development of the dough during fermentation depends on the capacity of the flour's gluten. The gluten must remain impervious to the carbon gas yet stretch enough to allow the dough to swell.

Before being put in the oven, the loaves are slashed, which makes the bread lighter and more appetizing.

## (c) Baking

From the time the loaves are put in the oven great changes take place in the dough: immediately the volume of the dough increases and a crust begins to form on the surface. This increase in size is explained by an intensification of the production of carbon gas, which is a result of the activation of the yeast. This increase of volume stops as soon as the temperature reaches 75°C/167°F. At this temperature, there is an inactivation of the 'raising' enzymes. At this point the alcohol begins to evaporate because of the heat. At around 90°C/194°F a difference becomes quite apparent between the crumb and crust. During the rest of the baking period, the temperatures of the crumb does not pass 100°C/212°F though the crust can reach 200°C/392°F. It is important to note that the crust begins to brown towards 110°C/230°F. The average oven temperature is generally 250°C/482°F.

During the first few minutes of baking, the gluten loses all the water it has mixed with during fermentation, and it ends by coagulating around 70°C/158°F. The water released during the coagulation diffuses into the dough forming steam which contributes to the dough's expansion. This water equally allows a better hydration of the starch. The enzyme responsible for the starch breakdown during fermentation is also impaired by heat (around 70°C/158°F). But we should note that for the short period after the dough is put in the oven, the enzyme continues to produce appreciable quantities of maltose.

The brown colour of the crust is a result of the caramelization of the sugars on the surface of the crust. Baking lasts from 15-30 minutes, according to the weight of each loaf. Contrary to the widespread use in most countries where bakers use moulds, the baking in France is done on trays.

When the bread is taken out of the oven it cools down slowly, the steam and carbon gas gradually exchanging places with the surrounding air, an exchange which happens through the crust. This is the 'sweating' period. This might make you think that it is a continuation of this process which accounts for bread going stale. But even in a very damp atmosphere bread can go stale. This happens because the starch within the bread gradually loses its initial suppleness and becomes hard and stiff. This retrogression of the starch can be stopped if the bread is reheated to 60°C/140°F. The starch again becomes supple and elastic and the bread 'fresh'. You can thus make use of this property when freezing bread and by reheating it in an ordinary oven. This practice is becoming more common amongst those who own a deep freezer. One should stress the fact that fresh bread is indigestible. Given that it has a certain elasticity, the digestive juices have a job to penetrate it. It is thus heavy to digest, especially if one eats it quickly without chewing it sufficiently. Fresh bread is not recommended for those who suffer dyspepsia or for elderly people. Stale bread is far more easily digested.

**Old-fashioned breadmaking:** the traditional method of 'three rises'.

—Take from the last batch of the evening 4 lb (2 kg) of dough – which is the main leaven.

—Make a ball with this and place it in a basket in a fresh and airy place in summer, temperate in winter.

—Leave it to rest for about 10 hours.

—Then add 17 lb (8 kg) of flour and enough for a firm consistency: this is the first leaven.

—Leave the leaven for 8 hours.

—Add another 17 lb (8 kg) of flour to the first leaven and sufficient water: this is the second leaven.

—Leave this for 1 hour.

—To the 34 lb (16 kg) of the second leaven, add enough flour and water to make 120 lb (54 kg) of dough: this is the third leaven.

—Leave this for 4 hours.

—Knead the dough adding the same weight of flour and water 120 lb (54 kg): you thus have 54 2 kg loaves which you should cook for 1 hour at 250°C/482°F.

## Making leaven for bread, for the first time

Mix: One cupful of wholemeal flour (from organically grown wheat).

—Half a cupful of pure water (bottled spring water).

—One teaspoonful of cold-pressed olive oil with an acidity lower than 0.5.

—One teaspoonful of acacia honey, pure and unheated.

—A large pinch of sea-salt.

The dough should have a moulding consistency and should not stick to your fingers.

Place the dough in an earthenware pot covered with a cotton cloth. Leave to rest for three days at a temperature of about 20°C/68°F. Your first leaven is ready: you can now make your first loaf.

## Making your first loaf

For four people for a week, mix:

—3 lb (1.5 kg) of wholemeal flour (organically grown wheat).

—1½ pints (750 ml) of pure water (bottled spring water).
—One tablespoonful of sea-salt.
—Your leaven (it has now rested three days, becoming soft, acidic and swollen).
—Mix and slowly knead these four ingredients (flour, water, salt, leaven) for at least 15 minutes, stretching and twisting the dough in all directions.
—Leave the dough to rest for 24 hours (taking care to put aside a ball of dough which you will use as leaven for your next batch).
—Make 4 loaves which you will bake in a hot oven after slashing the surface of the dough.
—Cook for about 30 minutes, waiting until a knife comes out clean.
—Don't eat the bread for several hours after baking.

## Two ways to use up stale bread

### Bread Soup

*Serves 4*

| Imperial/Metric | American |
|---|---|
| 10 oz (300g) of stale wholemeal bread | 10 ounces stale wholemeal bread |
| 4 pinches of sea salt | 4 pinches of sea salt |
| 6 cupsful of water | 7 cupsful water |
| 2 egg yolks | 2 egg yolks |
| ½ oz (15g) butter | 1 tablespoonful butter |

1. Cut the bread into thin slices and leave them to swell in the salted water for 5 minutes.

2. Heat over a low flame for 15 minutes stirring from time to time.

3. Remove from the heat.

4. Add the beaten egg yolks.

5. To serve add the butter.

# Le Pain Perdue (leftover bread)

*Serves 4*

**Imperial/Metric**
2 eggs
1 pint (570ml) milk
10oz (300g) stale wholemeal
  bread
Vegetable margarine
Honey

**American**
2 eggs
2½ cupsful milk
10 ounces stale wholemeal
  bread
Vegetable margarine
Honey

1. Beat the eggs and add the boiling milk.

2. Cut the bread into thin slices and soak them into the mixture.

3. Drain the bread and fry in the margarine.

4. Serve, spread with honey.

# 6.
# WHEATGERM

**The Synergy of the Natural Product**
Every book on dietetics waxes eloquent on the subject of
wheatgerm; 'a miracle food', 'irreplaceable', 'marvellous' . . .
And, in a book on cereals, we should give a particularly
important place to this product which is so hallowed by
dieticians. But we must first dispel some of the mysteries
which seem to surround wheatgerm.

Wheat germination, the importance of which was
recognized at the beginning of the century by Dr Carlton,
seems to liberate a vital energy latent in the grain. It is as if
the process of germination purifies the action of the
elements which make up the germ. This is the hidden
dimension which illustrates the advantages of the natural
product compared to the reactions which can be isolated
and synthesized in a laboratory. The great error of
chemotherapy is to forget the laws of nature, the complex
but necessary ordering of natural elements, to remove
certain useful bodies from their biological environment.
For there is more to the biological sum total of a living
body than the simple addition of its constituent parts. In
biology $2 + 2$ never makes 4, but always a little more than
4, and it is at this level that phenomena, sometimes
beyond human investigation, create a synergy, rich in
supplementary power.

The extraordinary biological value of wheatgerm is not
only explained by the total properties of its elements, but
by the interaction between its elements, by the harmony
which presides over all, the symbiosis of its powers. Even
if it escapes our understanding, we should recognize the
synergy which characterizes wheat germination.

During this germination, enzymes intervene, catalysts whose action facilitates the body's assimilation of the principal elements of the germ: mineral salts (calcium, potassium, magnesium, sodium, phosphorus, iron and sulphur), trace-elements (magnesium, nickel, copper, bromium, aluminium, zinc and iodine), vitamins (A, $B_1$, $B_2$, $B_3$ (PP), $B_5$, $B_6$, C, D, and E) proteins (essential amino acids: arginine, cystine, histidine, isoleucine, leucine, lysine, methionine, phenylalaline, threonine, tryptophane and valine) and lipides (polyunsaturated fatty acids; oleic, palmitoleic, linoleic, linolenic and saturated: stearic and palmitic). Amongst these enzymes are: amylase, glutenase, cystase, peroxydase, carboxylase and lipase.

We see, when stating the constituents of wheatgerm, its quite exceptional richness. It is a highly unusual product, all the more because the process of germination enriches the content of the germ in essential elements, as the following table, a comparison of the percentages of principal mineral salts and vitamins in the composition of the germ and the percentages of the same elements in white bread and wholemeal bread, indicates. An examination of this table shows the consequences of germination.

| Mineral Salts in mg | White Bread | Wholemeal Bread | Wheatgerm |
|---|---|---|---|
| Calcium | 20 | 40 | 90 |
| Magnesium | 10 | 130 | 400 |
| Phosphorus | 90 | 390 | 1100 |
| Potassium | 100 | 290 | 800 |

| Vitamins in mg | White Bread | Wholemeal Bread | Wheatgerm |
|---|---|---|---|
| A | 0 | 0 | 0.4 |
| $B_1$ | 0.07 | 0.60 | 1 |
| $B_2$ | 0.05 | 0.10 | 2.5 |
| $B_3$ (PP) | 0.7 | 4 | 5 |
| $B_5$ | 0.4 | 0.8 | 2.1 |
| $B_6$ | 0.2 | 0.9 | 3 |
| C | 0 | 0 | 1 |
| D | 0 | 0 | 0.07 |
| E | 0.2 | 3 | 27 |

## The Properties of Wheatgerm

After a quick run through of the properties of each of its components, it is easy to see the astonishing therapeutic properties of wheatgerm.

*Note:* The quantities of mineral salts and the vitamins indicated in the table (in milligrams) are given for 100

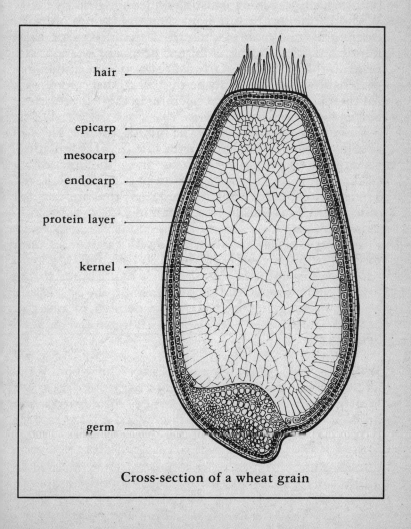

**Cross-section of a wheat grain**

grams of product (white bread, wholemeal bread or wheatgerm).

## (a) Mineral Salts

**Calcium:** Almost all the calcium (3 lb/1.5 kg in an adult) is concentrated in the skeleton: it is the mineral for bone-building, responsible for strong bones and healthy teeth. Without calcium in sufficient quantity, bones are soft, fragile and vulnerable to fractures. Calcium is particularly important for growing children, pregnant women who need double the calcium to ensure good bone building in their babies, and for elderly people, who must compensate for their perceptible loss of calcium. In these three cases a better diet is called for.

Calcium is also present in the blood, where it plays an absolutely fundamental role in the process of coagulation. Calcium exercises a regulating action on the functioning of the nerves and muscles, and in particular, the heart. The assimilation of calcium requires the presence of phosphorus, magnesium and vitamin D.

**Phosphorus:** is always associated with calcium, as their metabolisms are closely linked. Like calcium, it plays a part in bone building. Moreover, it contributes to the formation and regeneration of the brain, the nerves and all organs in general. It is recommended in cases of physical and intellectual asthenia, fatigue and mineral deficiency.

**Sodium:** is the basic element of intercellular liquid, the liquid in which all the cells of the body bathe. It participates in all functions of the body, and we need to keep replacing it because it is eliminated in urine, when it is hot and we sweat, and when we exert ourselves.

**Potassium:** As opposed to sodium, it is contained in intracellular liquid, and its presence is indispensable to the subtle changes which take place within each cell of our body. A good balance of sodium and potassium is a primary condition for good health.

Potassium is a muscle stimulant, and in particular, a toner for the heart.

**Magnesium:** Is the object of endless research because of the relationship which seems to exist between the development of cancer and other wasting diseases and our intake of magnesium (because of the excessive refining of our food). We are all, some more than others, according to our diets, lacking in magnesium. This mineral plays a part in the metabolism of calcium and phosphorus, although we still do not understand how it is assimilated into the body and plays a part in the metabolism of calcium and phosphorus. Magnesium reinforces the body's natural defences against infection, microbes, viruses and – so it seems – against cancer. Preventing senility, it slows down the ageing process of the body.

**Iron:** This is the principal constituent of haemoglobin in the blood, thanks to which it can fulfil its function of the fixation and transport of oxygen through the body. If the iron content of haemoglobin is insufficient, anaemia follows, and the body's tissues are badly oxygenated: a general poisoning of the body develops.

**Sulphur:** This has a beneficial action on body tissues, in particular the skin and the arteries which it softens, fighting against premature ageing. It also allows the liver to filter away body toxins. Finally, it is beneficial to rheumatism, to those suffering with arthritis, arthrosis and gout.

All these mineral salts, which contribute to the harmonious function of the human body, are present in wheatgerm, in compatible proportions with the body's demands.

Their interaction in the framework of certain complex, cellular metabolisms underlines the superiority of the natural product – in its rich totality – in relation to the elements which allow their assimilation. But if these mineral salts, found in important quantities in the body, have been known for a long time, one can count today other minerals present in infinitesimal quantities, in some cases traces of these minerals only: trace-elements (etymologically speaking). One should pay particular attention to these minerals, for everything indicates that their action – inversely proportional to their quantitative importance – regularizes cellular functions; organic, endocrine and nerve. Each of these elements, even if present in tiny traces, plays a precise role, often as catalyst for another reaction.

## (b) The Trace-Elements

**Manganese:** This regularizes the glandular system, improves the functioning of the liver and kidneys and helps the body assimilate the B group of vitamins. Moreover, it is known for its effect in cases of allergy; asthma, nettle-rash, eczema, etc.

**Nickel:** Stimulating pancreatic functions, nickel is recommended for diabetes. It is also thought that it checks the development of cancerous cells.

**Copper:** Facilitating the fixation of iron, it plays a part in the constitution of haemaglobin in the blood. It also acts in blood coagulation. We are well aware of its valuable properties in fighting infections and viruses.

   In cases of flu, the body mobilizes its reserves of copper to fight against the attack of microbes. Finally, copper works against tissue degeneration.

**Bromide:** This is a powerful sedative for the nervous system particularly recommended for insomniacs.

**Aluminium:** Just a trace of aluminium has a marked calming effect on the equilibrium of highly-strung people.

**Zinc:** Is a stimulant of the hypophytic and sexual glands. By its action on the pancreas, it improves the secretion of insulin, which makes it indispensable to diabetics.

**Iodine:** The chief importance of iodine is in its role in the functioning of the thyroid. A deficiency in iodine causes numerous problems in the body.

Iodine acts against hypertension, obesity and rheumatism. It is a precious mineral for growing children and for the ailments of elderly people.

As with mineral salts, there is often an interdependence between trace elements, whose actions are complementary. The action of the trace-elements in wheatgerm is intensified by the synergy of their natural environment.

## (c) Vitamins

**Vitamin A:** This is the anti-infection vitamin, which reinforces the body's natural defences against the attack of microbes.

Through its regenerative action on the epithelium, it keeps the skin supple and young. It facilitates the healing of wounds and helps nocturnal vision. It is an indispensable vitamin for growing children.

**Vitamin $B_1$:** Helps the assimilation and metabolism of glucids in each cell. Its work is thus fundamental to the whole body. It assists in resisting neuritis by helping to guard against inflammation of the nervous system, and the serious problems that follow.

Lack of vitamin $B_1$ leads to beriberi. Beriberi is a very formidable illness characterized by lack of vitamins through refined cereals (polished white rice). This is another example of our dietary errors by favouring white cereals over whole cereals.

Beriberi is a 'historic' illness in as much as its study at the beginning of the twentieth century discovered vitamins. Its effects are dramatic: polyneuritis, cardiovascular problems, oedemas, paralysis, asphyxia leading to death. Beriberi was prevalent in all South East Asia at the beginning of the century, and also in parts of Latin America and Africa. After a while people realized that its prevalence was directly linked to the abandonment of whole cereals and to the consumption of white, polished, refined, denatured rice.

It is only following much research – by watching the people who eat brown rice and observing that they do not suffer from the disease – that one grasps this relationship between the refining of rice and beriberi. These studies directed the biologists' investigation towards the composition of the peripheral layers of the grain, and they resulted in the discovery made by Professor Funk, who isolated vitamin $B_1$ and coined the word vitamin. This event justified to him alone the total and definitive abandonment of refined cereals and the necessary and exclusive choice henceforth of whole cereals. Without going to the extreme of beriberi, those moderately lacking in vitamin $B_1$ are neurasthenic, depressive, feeble, they have dizzy spells and lack appetite.

**Vitamin $B_2$:** This operates on a cellular level in the metabolism of sugars, amino acids and lipids. It aids vision in darkness and contributes to the health of the skin. It also stimulates the secretion of insulin. A lack of vitamin $B_2$ leads to a deterioration of eyesight, conjunctivitis, cracks in the skin and general dry skin.

**Vitamin $B_3$ or Vitamin PP** (Pellagra Preventive): Vitamin $B_3$ keeps the skin in perfect health, balances the nervous system and preserves the suppleness of blood vessels. It helps in the assimilation of sugars in each cell of the body.

Lack of vitamin $B_3$ leads to pellagra, a malady endemic in those countries where maize and millet are the basic diet (as opposed to other cereals, these two cereals do not contain vitamin $B_3$). Pellagra is characterized by digestive and nervous problems.

**Vitamin $B_5$:** This vitamin is really the pivot of the biological changes which happen in body tissues, organs and cells. It allows a good utilization of food and production of energy. Lack of vitamin $B_5$ leads to fatigue, insomnia and digestive problems.

**Vitamin $B_6$:** Vitamin $B_6$ enables the breakdown of amino acids: an increase of proteins in the diet must thus be accompanied by an increase in $B_6$. It helps in the formation of white blood cells. It keeps the skin healthy: a lack of vitamin $B_6$ leads to skin diseases, and eczema around the mouth and eyes.

**Vitamin C:** Guards the body against all sorts of microbes, for example colds and flu. In effect, it stimulates the formation of antibodies. It helps in the assimilation of sugars and the formation of muscular glycogen. It wards off haemorrhages and protects the body against the stress of modern life.

A lack of vitamin C leads to scurvy, which is recognized by the occurrence of serious problems in the stomach, intestines and blood vessels.

**Vitamin D:** Its role is absolutely fundamental: it allows the assimilation of calcium by facilitating its passage in the blood through the intestinal wall. It is vitamin D which makes the fixation of calcium possible. Its lack leads to a lack of calcium and rickets.

**Vitamin E:** This stimulates the sexual glands, helps the reproductive functions and fertility. It acts on the hyperphyse and regularizes certain nervous activities. Vitamin E deficiency leads to reproductive organ problems, in particular to impotence.

After this description of the various remarkable properties of the numerous elements which make up wheatgerm, one might be tempted to think of it as a panacea of curing all ills. Without going that far we must insist on one point:

Wheatgerm is the natural alimentary product which is most rich in: magnesium, phosphorus, vitamins of the B group and E.

Also, we advise the regular consumption of wheatgerm in the following cases: breast-feeding, anaemia, growth, calcium deficiency, mineral deficiency, fatigue, frigidity, obesity, impotence, lymphatism, rickets, scurvy, old age, tuberculosis.

Moreover, as wheatgerm is so rich in magnesium, we recommend its regular consumption to avoid a deficiency that we have good reason today to associate with cancer.

One can find wheatgerm and products containing wheatgerm (rusks, biscuits, cereals, etc.) in all health food shops.

Finally, we cannot end this chapter without giving you the recipe to germinate wheat yourself.

### How to Germinate Wheat
It is in your interest to buy wheat to germinate yourself in addition to eating wheatgerm. It is very easy to germinate wheat:

The cycle of germination lasts 3 days. You must buy organic wheat from a health food shop.

Day 1 – Wash the wheat in cold water and put in a large dish (or plate), and cover with pure water (we recommend bottled water).

Day 2 – Wash the wheat and replace it in the dish, but leave it damp, without covering it with water this time.

Day 3 – A white dot should appear on each grain: this is the germ. The germinated wheat is now ready to be eaten.

We advise eating two tablespoonsful of fresh wheat germ with your midday meal, chewing it carefully (two teaspoonsful for babies). Repeat this daily for three weeks.

# 7.
# THE DIETARY AND THERAPEUTIC PROPERTIES OF BRAN

'The world is illogical: servants throw bran to the pigs even though it includes more healthy and nutritious substances than flour.'

Sebastien Kneipp 1886

'The regular consumption of cereals rich in bran is the best remedy for constipation.'

Joseph Favrichon 1903

## The Rediscovery of Bran

Sebastien Kneipp in the nineteenth century, and Joseph Favrichon, at the beginning of the twentieth century, looked at the evidence for the remarkable dietetic and therapeutic properties of bran: the quotations we have chosen, at the head of this chapter, illustrate this interest for a product which for more than fifty years was ignored; discredited by nutritionists and despised by doctors. This has been a barren age for bran with no place for such a product in contemporary food consumption marked by a rise in living standards and a rejection of old dietary habits. Manufacturers became masters of the market, preoccupied solely by the idea of yield and profitability and above all . . . refining. Refine, refine, refine . . . everything must be white: flour, bread, sugar, salt . . . and all the world is content, even the dieticians!

Professor Tremolieres wrote: 'Bran is an agent of disassimilation, for it accelerates the intestinal journey, which limits the duration of the action of the digestive juices.'

Why do we use bran? It is not broken down by the digestive

juices and therefore cannot be absorbed by the organism, so it accelerates the progress of waste matter through the body. Without the presence of bran the waste products remain in the digestive tract for much longer and it is during this time that diseases such as cancer, diabetes, cardiovascular problems, dental decay, rheumatism, constipation, and obesity can take hold. In the past the scapegoat has been pollution, central to our ecologically disrespectful civilization. Then, abruptly, a few years ago attitudes changed, opinions altered. Scorned for half a century, bran is suddenly in favour again, adored by the High Priests of medicine and dietetics. Since the Bichat talks of 1974, first gastroenterologists and then general practitioners have come round to the way of thinking that bran is no longer a curse. In the health food shops there are constant streams of people who come in to buy a packet of bran, bran rusks, bran biscuits . . . At one point the Favrichon company, who had abandoned making granulated cereals with bran a long time ago, put their product back on the market. 'Granosson' had its hour of glory at the beginning of this century when it was exported all over the world, before its decline set in.

Bran is now everywhere. It appears in scientific reviews; publicity has made it the symbol of success. The craze for bran has surprised even the sceptics. We cannot discuss cereals without giving bran top priority. It deserves this honour because of its properties in the realm of gastro-intestinal pathology, and in particular, constipation. We will look closely at the relationship between bran and illness – including the disease which has spawned the largest number of scientific investigations in our era – cancer. In particular, we will look at cancer of the colon.

### What is Bran?
Bran is the outer coating of cereal grains, called the pericarp. It is formed by three membranes, the epicarp, the mesocarp and endocarp. Crushed during the milling of the whole grain, it can be separated by sifting, from the other products of milling (proteinic layer, germ and kernel). Brown in colour, soft to the touch (despite its

appearance), it is composed of two vegetable fibres: cellulose and hemicellulose. These fibres are complex glucids called polysaccharides; glucids endowed with a remarkable property. They are not digestible by one single digestive juice, as opposed to all the other carbohydrates which are transformed into more easily assimilated sugars by digestive enzymes. In effect, man does not secrete the single enzyme capable of digesting these vegetable fibres: cellulose. Only ruminants, which do have this enzyme are capable of digesting bran.

It is precisely because man cannot digest bran that it has exceptional dietetic and therapeutic properties. That might seem paradoxical at first sight, but the whole of this chapter will explain this apparent contradiction.

Bran, not broken down during digestion, is a precious ballast: it increases the volume of the body's waste matter, and we will see how this property is useful to the harmonious functioning of the digestive tube. Moreover, endowed with a considerable hydroscopic power (the power to absorb water), it increases the volume of waste considerably in the course of its travels down the digestive tract. Bran soaks up and absorbs twenty times its own volume of water! It swells up enormously and this allows it to shift the food in the upper part of the digestive tract, and the by-products of digestion in the intestine.

We shall see in the section devoted to constipation that this quality constitutes a real benefit to our health. It is in fact in the pathology of constipation that bran has its most judicious use.

## Constipation
In France in 1978, one woman out of two suffered from constipation, and one man out of five. One can say that this complaint is responsible for numerous other problems!

Constipated people have great digestive problems, suffer from flatulence, stomach distention, furred tongues and bad breath!
—They find it difficult to wake in the morning.
—They are permanently tired.
—They are depressive and asthenic.

—They sleep badly and are restless.
—They suffer from frequent migraines.
—They have skin problems, e.g. acne and eczema.
—They are sensitive to the cold.
—They frequently have rheumatism in their joints.

Constipation leads to a self-intoxication as a result of the permeability of the intestinal mucus to the multi-millions of toxic substances which proliferate in the intestine. The intestinal mucus cannot stand up to the invasion of these microbes and when the kidneys and liver – which form the second barrier against the invasion of toxins – are themselves taken over, the whole body is invaded by deadly poisons carried by the blood.

*What are the causes of constipation?*
The number one factor seems to be too much sitting which goes hand in hand with urbanization and the rise in the standard of living: people are just not getting enough exercise, are not walking enough, and are getting bad habits which they will live to regret, in particular, automatically getting into the car even to go just a very short distance.

The stress of contemporary living does not help things: people are continually rushed, excited and overwrought, which leads to many alimentary problems: they eat no matter where, no matter what, no matter how, rushing, without chewing properly, surrounded by noise, watching the television, reading.

In the whirlwind which surrounds them, they get into the habit of resisting the 'calls of nature', each day a little longer. Finally, a diet increasingly rich in sugars, in fats and meat seems to be of extreme importance to the development of constipation. A meat starter, followed by steak and chips, a bought ice cream generously covered with whipped cream to finish, the whole lot accompanied by white bread. Does this menu ring a bell with you? In such a meal, you cannot find much ·fibre! It's hardly surprising that constipation is the result.

*Constipation is really the characteristic ailment of our civilization*
Doctors Burkitt and Trowell have worked together for twenty years in Uganda and they have reached the following conclusion:

> The lack of cellulosic matter in the daily diet of the populations of industrial countries results in a cascade of illnesses, including constipation.

The observations of these doctors are based on a comparison of the English diet with the Ugandan diet. In the former, a diet poor in vegetable fibre: in the latter a diet rich in fibre; whole cereals, green vegetables, etc. A comparison of the transit time of food through the body gives us a valuable indication: 24 hours for the Ugandans (average time between the evacuation of stools) and 72 hours for the English! Thus the predominant role of bran and other vegetable fibres (those of green vegetables) in the acceleration of food's intestinal journey is proved. Of course, those who are constipated will be better if they take more exercise, walk and relax. But they must not ignore the advice of the dietician who advises us to reintroduce an essential food into our diet: bran.

## The Digestive System and its Function

> 'Having spent a long time studying the non-digestible parts of our foods, we have discovered that several modern illnesses are provoked by insufficient consumption of them.'

> Dr Denis Burkitt (1977)

Man's digestive tube – from mouth to anus – has a structure adapted to its functions: each of its parts possesses anatomic and physiological properties relating to the role which it plays. A better understanding of the developments which are to follow as regards the value of bran in diet, and in gastro-intestinal pathology, necessitates a concise reminder of several elementary ideas on the function of the digestive tube.

The first fundamental transformation takes place in the *mouth*, where the food is ground and saturated with saliva. This thorough chewing and thorough salivation, increases the digestibility of food. It is at this point that the first digestive enzyme acts, an amylase called ptyaline, which undertakes the breakdown of complex sugars like starch.

The ball of food resulting from this first digestive function crosses the pharynx and enters the oesophagus where muscular contractions push it along the twenty to thirty centimetres of its journey. The passage of the alimentary ball in the upper part of the stomach is regularized by the movements of the sphincter, a circular muscle at the entrance to the stomach.

In the *stomach*, contractions mix and homogenize the food ball, at the same time activating gastric secretions:
—mucus, which covers the internal walls of the stomach with a protective film which preserves them from corrosion.
—hydrochloric acid, which acidifies the food ball, and allows a molecular dissociation of the proteins.
— gastric juices, which comprise three diastases: pepsinogen, transformed to pepsin by hydrochloric acid, which leads to the initial breakdown of proteins: gastric lipase, responsible for an elementary transformation of fats; and rennet which helps to coagulate milk.

The food ball, mixed by the stomach and impregnated with gastric juices, then moves into the intestine, where it is then called chyme. The entry of the chyme into the intestine is controlled by another sphincter, the pylore, which adapts its contractions to the needs of the intestine, gradually evacuating its contents.

From the exit of the stomach, the chyme moves on to the *duodenum* where the choledoc canal, which starts in the liver, and the Wirsung canal emerge. The Wirsung canal arises in the pancreas. In the duodenum the chyme is impregnated with secretions which escape from these canals: bile secreted by the liver and pancreatic juice by the pancreas.

*Bile* gets its yellow colour from pigments which have no digestive action, and these pigments come from the breakdown of old red blood corpuscles. Bile does not contain digestive enzymes, but it contributes to the action of those of the pancreatic juice. It is principally made up of salts which play a very important role in digestion; the bile salts neutralize the acidity of the chyme, emulsioning the lipids in tiny particles, which the pancreatic enzymes can more easily attack, and most importantly, it accelerates the peristaltic movements of the intestine. It also contributes to the cleansing of intestinal flora by having an antiseptic action (hepatitis sufferers, whose bile secretion is insufficient, are generally constipated, and their stools have a nauseous smell).

*Pancreatic Juice* as opposed to bile, contains three digestive enzymes: pancreatic amylase, which continues the breakdown of the glucids which was begun by the ptyaline in the saliva; pancreatic lipase, which reduces fat molecules to allow their ultimate absorption by the intestinal mucus; trypsinogene, which activates the process of digesting protein.

At the beginning of the duodenum, the enzymatic breakdown of food particles is almost total. The chyme then enters the small intestine.

*The Small Intestine* is about 7 metres long, but the surface of contact between the mucus and chyme is infinitely greater, because this surface is made up of innumerable pleats, intestinal villosities. It is at this point in the small intestine that the majority of products of digestion are absorbed by the body, thanks to the minute blood vessels which line the villosities. This absorption is preceded by a final enzyme attack within the small intestine. Absorbed by the intestine's mucus, the molecules resulting from digestion are collected by blood vessels which effect their final assimilation by the body.

The small intestine ends with a swelling called the caecum, which is where the appendix is found. The food ball, now chyme from the stomach, changes its name again to faecal bolus, at this point, before it enters the large intestine.

*The Large Intestine*, about 1.5 metres long, is most usually called the colon. It has three sections: an upper part, at the exit of the caecum which is called the ascendant colon. This bends near the liver to cross the abdomen from right to left: this is the transverse colon. After the second bend, it redescends towards the rectum, becoming the descendent colon. It is in the rectum, which is about 15 centimetres long, that the faecal matter collects before expulsion through the third sphincter of the digestive tube: the anus.

The only digestive 'operations' which happen in the colon are the absorption of water and certain mineral salts (calcium) by the intestinal mucus.

The progress of the faecal bolus in the colon happens at a speed proportional to the quantity of waste arising from the digestion of the food ball: these waste products are the non-digestible parts of the food we eat, made up of vegetable fibre, and in particular, bran. The larger the volume of these wastes (i.e., the more fibre there is in the food) the quicker the progress of the faecal bolus. A small faecal bolus moves slowly in the colon: this is the state of constipation. On the other hand, a large faecal bolus, rich in cellulose and soaked in water, moves rapidly in the colon. We will see why.

The faecal bolus moves in the colon as a result of a process called segmentation. This process is quite simple. The colon is surrounded by circular muscles which contract to advance the faecal bolus. These contractions become closer and closer together in the following way: when two consecutive circular muscles press the colon, they isolate the part of the colon between, forming a pocket which traps a certain amount of faecal matter. By contracting again, the muscles increase the pressure inside this pocket until the part of the colon below the pocket dilates as a result of this pressure. The faecal matter moves along a few centimetres, and a new pocket forms under the pressure of two new circular muscles . . . and so on. This movement strangely resembles that of a centipede crawling along the ground! The speed of progress in the colon largely depends on the volume of bran in the faecal matter: in effect, when the faecal matter is large in volume, the circular muscles pressing on the walls of the colon can exercise their pressure easily and regularly. If the volume of faecal matter is small, the action of the muscles is slow because they exert a far less intense pressure, and the faecal bolus moves very slowly: this is a case of constipation.

## BRAN AND CONSTIPATION

**Bran in diets.** We have seen earlier on – because it is not broken down during digestion and that it is found intact in faeces – that bran increases quite considerably the volume of the faecal matter, in particular thanks to its astonishing property of absorbing water which causes it to swell; it can increase its volume up to twenty times. Thus, not only does waste rich in bran have a far larger volume, it is also softer than that resulting from a diet lacking in bran. It is consequently easy to understand why faeces move more quickly in the colon when it is soft, because the circular muscles responsible for its movement have a far easier job.

To sum up, a diet poor in bran (and vegetable fibres in general) produces a small and dry faecal bolus which the intestinal muscles have difficulty moving. A diet rich in bran and other vegetable fibres, produces large and soft faeces which facilatates easy and rapid movement. The introduction of bran into the diet is really the best thing in the world for constipation.

**The evacuation of stools.** The rectum takes far more time to fill when waste matter is small. The need to 'go' is thus only spasmodic, very irregular, infrequent and not urgent. Certain people who suffer from constipation, only open their bowels once a week! Their systems are completely poisoned and their health is obviously seriously impaired.

On the other hand, in cases where diet is rich in bran, the rectum fills swiftly, and 'the call of nature' is regular, frequent and imperative! The expulsion of stools by dilation of the anal sphincter, happens normally, the opposite of the expulsion of constipated stools, which takes a great deal of effort and tends to cause haemorrhoids.

**Laxatives.** We cannot end this section on constipation without emphatically condemning all synthetic laxitives. Chemical laxitives are amongst the most frequently prescribed medications, along with sleeping pills, tranquillizers and diuretics.

Of course, the use of a laxative has immediate and spectacular effects, but an abuse of them leads to absolute dependence. Those who regularly take laxatives, need to do so all their lives, they can no longer get by without them. Chemical laxatives seriously irritate the intestinal mucus, which provokes very distressing colitis and all sorts of chronic intestinal ailments. There are numerous drawbacks to taking laxatives, when really all you should do is readjust your diet.

Those who still wish to use laxatives should buy them from health food shops to be sure that they are made entirely from natural products. Several types exist in the form of tablets or capsules. In general, they are made totally from vegetables; rhubarb is the one most frequently used. Natural vegetable laxatives do not create dependence and do not irritate the intestines.

### Bran in Gastro-intestinal Pathology

Bran not only helps in the pathology of constipation: its action has great effects in the cure – or prevention – of other gastro-intestinal problems: diverticulitis, appendicitis, haemorrhoids, and cancer of the intestine (we will be looking at this particular aspect separately).

**Diverticulitis.** This ailment, unknown at the beginning of the century, is today extremely common, affecting the large intestine. It is, along with constipation, a character-

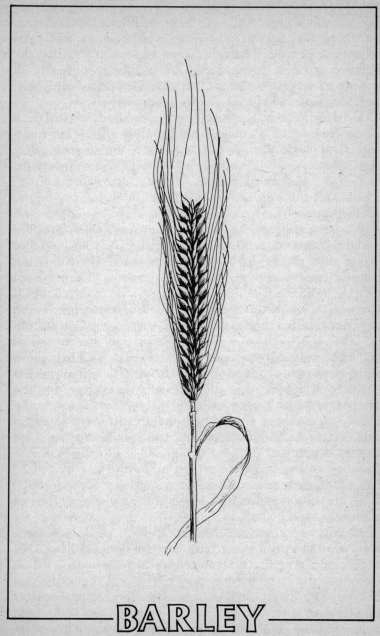

BARLEY

istic problem of our age of urbanization, of sitting down and of a diet lacking in bran and fibre. Burkitt and Trowell observed that this problem was extremely rare in rural Ugandan people. Other studies have shown that the Japanese, too, do not have this illness, though it does occur amongst Japanese emigrants in the USA, who have adopted the American way of life as well as the diet.

Those who do not suffer from constipation, whose faeces move freely in the colon, because they are voluminous, soft and watery, do not suffer from diverticulitis, the characteristic symptoms of which are intestinal swellings, abdominal pains, nausea, hard and painful stools and the unpleasant feeling of never emptying the rectum completely.

We have seen how faecal matter moves down the colon by simultaneous action of the circular muscles trapping waste in a pocket, which empties when the following muscle exerts enough pressure to allow the previous muscle to dilate. These small pockets, the basic wheels of the peristaltic movement, are called diverticules. When the faecal bolus is hard and small, there is not enough to grip for the circular muscles to exercise pressure. The segmentation is thus poor and irregular, and the colon walls have a tendency to shrink: the peristaltic movement is then affected, and the pressure exercised by the circular muscles becomes progressively less intense and lazy. The pressure finally becomes insufficient to expel the matter in the diverticules: the upper muscle is no longer capable of dilating the muscle below. Then, it is the wall of the intestine which takes the strain, and all sorts of hernias occur along the way: this is diverticulitis. The intestinal wall begins to resemble an old inner tube of a bicycle tyre!

As you know, inner tubes which have been used for a long time sometimes explode. It's the same in the case of the colon, and the consequences are dramatic. When diverticulitis has become really bad and segmentation becomes really feeble and irregular, the stagnant matter infects the intestinal mucus, which in turn makes it vulnerable by reducing its capacity for resisting pressure.

This is shown by the difficulty the circular muscles have in moving the waste. Thus, sometimes perforations of the intestinal wall occur and the end can be an explosion . . . this is peritonitis, which is fatal in every case if an operation does not happen immediately to close the broken wall. Luckily such complications of diverticulitis are relatively rare. But one should be aware of the their existence, or we would not be doing those who suffer from constipation a service.

The answer is so simple! The introduction of bran into your diet ensures that normal intestinal functions are re-established, particularly in those people whose bowel movements are particularly lazy.

**Appendicitis.** The appendix operation is the most frequent abdominal operation in the western world, and as with diverticulitis, inflammation of the appendix leads sometimes to peritonitis and death unless immediate surgical action is taken.

Dr Stanway has given worrying examples which prove the direct relationship between inflammation of the appendix and lack of vegetable fibre in the diet, in particular a lack of bran: this problem was very rare in the last century but changes in our eating habits in the western world have led to its marked increase. Stanway made these observations about the American army around 1920: at that time appendicitis was unknown to Negroes, but with their enrollment in the army and ensuing diet, it became apparent that Negro soldiers, during the Second World War, began to experience appendicitis. The same thing happened to students from the Congo who went to Belgium to study. He made these observations in Great Britain before the war: the children who went to boarding school, the offspring of the upper middle classes, suffered from appendicitis, but those children in orphanages, where the diet was courser than at fashionable schools, in other words richer in bran and other vegetable fibres, appendicitis was unknown. The same observation can be made in comparing the diets of country people who live in rural villages without access to huge supermarkets, and city dwellers.

The appendix is a small narrow organ emerging in the caecum, the hinged area between the small intestine and the colon, on the right side of the abdomen. Appendicitis happens when the entrance to the appendix becomes clogged, provoking an inflammation which can lead to an explosion and to the dispersal of infection outside the intestine. As one would imagine, the principal obstruction in the appendix is faecal matter, when it is hard and dry as a result of low content in undigested elements (bran, vegetable fibre in general). The faeces of those people who have little fibre in their diet are hard and dry. It only takes one of these little balls to block the entrance to the appendix and trouble ensues.

Regular consumption of bran prevents inflammation of the appendix and the serious crises which can happen as a result, particularly peritonitis.

**Haemorrhoids.** Haemorrhoids are varicose veins in the lining of the rectum (internal haemorrhoids), or on the inside of the anus (external haemorrhoids). These veinous dilations lead to burning sensation, pruritis and to bleeding during defecation. It is all the more disheartening because haemorrhoids are irreversible (they never disappear once formed, except by operation) and the bleeding which is characteristic of haemorrhoids can prevent an early diagnosis of cancer of the colon. Because those who suffer from haemorrhoids are used to bleeding when they open their bowels, they do not take much notice, and cancer of the colon can develop unheeded.

The causes of haemorrhoids are hereditary, too much sitting down, malfunction of the liver, certain pathological states due to poor blood circulation and, of course, a long period of constipation.

Long and intense efforts to expel stools from the rectum seriously affect the numerous vessels which cover the mucosa and which surround the anus. These veins swell up, bloat and sometimes tear. In effect, they are put under too much pressure. All the pressure is put on the anal area when someone who is constipated tries desperately to expel stools which are hard and dry, a result of a diet poor in bran.

The consumption of bran leads to the formation of soft stools which are easy to expel because they are flexible due to their capacity to absorb water. It is in the interests of those who suffer from haemorrhoids to eat bran on a regular basis. Even those who do not have piles should take small quantities of bran as a preventive measure.

## Gall Stones and Bran

The gall bladder is a small organ in the form of a bag situated beneath the liver, which acts as a reservoir for bile secreted by the liver. The gall bladder stores up bile in proportion to its secretion and only 'delivers' it when needed, that is to say during digestion after a meal. The bile runs along the choledoc canal to rejoin the duodenum. However, in our present age of overeating, of obesity and hypercholesterol, this function no longer happens as naturally as in times gone by when less sugar, less fat, less meat and less refined, denatured foods were eaten.

Gall stones (crystalized mineral salts) interfere with the normal functioning of the gall bladder in countless western people who eat a poor diet. Cholecystectomy (the removal of the gall bladder) is the most frequent of voluntary operations (not imperative like appendicitis): more than a million gall bladders are taken out each year in the western world, of which 400,000 are in the USA!

What is most amazing is that most of the time the operation does not seem to make a lot of difference – as far as the pathological hepatic state is concerned a few months later. But it is impossible then to operate: there is then no gall bladder to remove! An operation in most cases can be avoided through dietary reform. A visit to a health food shop will give a few basic ideas on how to improve this condition. Begin the day with a glass of black radish juice . . . but we are straying from the point. What can bran do for the gall bladder?

The biliary lithiase is directly linked to the quantity of bran in the diet. We know that bran accelerates the intestinal journey of food and consequently reduces the time which the faecal bolus is in contact with the intestinal mucosa, through which certain products penetrate, which results in the ultimate digestion of food. Amongst these substances which take advantage of the permeability of the intestinal wall to join the blood vessels which carry them throughout the body, is a biliary acid, deoxycholic acid. Reaching the liver by the vein which carries all the substances absorbed by the intestinal mucosa, deoxycholic acid counteracts the hepatic synthesis of cheno-deoxycholic acid. This acid is extremely valuable: it neutralizes the formation of biliary lithiase, an accumulation of which leads to the formation of stones in the gall bladder. By accelerating the intestinal journey (by ensuring that bran is present in one's diet), the diffusion of deoxycholic acid through the colon wall is reduced, and as a result of this the liver can work more effectively against the formation of gall stones.

If you suffer from gall stones, eat bran!

## Slimming and Bran

We have shown in the introduction to this chapter, that the principal argument allowing nutritionists to discredit bran, rests on the fact that bran, by reducing the intestinal journey, limits the time during which gastric juices can act. Professor Tremolieres speaks of 'bran as an agent of disassimilation'.

However, on the contrary, the introduction of bran into the diet of the majority of overfed westerners has a beneficial action: by accelerating the intestinal journey, it limits the time for absorbing the substances resulting from the digestion of the food ball: when faeces remain for several days in the intestine, assimilation is effectively longer. We will see that this not only is probably the cause of cancer of the colon, but also that prolonged stagnation of these matters in the intestine leads to a greater absorption of calories: this increase can be as much as 30 per cent, and these are the supplementary calories which overweight people can do without.

Moreover, and this is important, bran – and all fibre – induces the feeling of having eaten enough: (wholemeal bread more than white, brown rice more then white rice) and we eat less because the feeling of fullness is more quickly reached.

In short . . . to lose weight, eat bran.

## Bran and Cancer of the Colon

Cancer research in Lyons in 1977 which looked into the physiopathology of cancers of the digestive tract and the eventual means of their prevention, came to the following conclusion: 80 per cent of cancers are environmentally linked within our bodies to the nature of our diet. An early diagnosis from suspicion of pre-cancerous areas, and a transformation of diet, can lead to prevention of cancers of the digestive tube.

We will not go into cancer of the oesophagus here, which is directly linked with the abusive consumption of alcohol, nor cancer of the stomach and liver, in which the value of bran has a relatively minor influence; but we will concentrate on cancer of the large intestine (there are very few cancers of the small intestine). The development of cancer of the colon seems directly associated with the quantity of non-digestible products present during the journey of the alimentary bolus through the stomach, duodenum and small intestine. The frequency of cancer of the colon is inversely proportional to the richness of bran in the diet.

Cancer of the large intestine is the next most common cancer to cancer of the breast in women and lung cancer in men. Around 25000 French people have cancer of the colon each year and 15000 die as a result. These are worrying figures, all the more so because early diagnosis is relatively easy in this type of cancer and a natural diet which is healthy and well-balanced, can prevent it to a large extent. It is true that we must all look to our eating habits.

The diagnosis of cancer of the colon is possible in the majority of cases. In fact, this cancer develops very slowly, over a period of ten to twenty years sometimes. As Professor Rene Lambert put it (Director of the Institute of Digestive Pathology in Lyons, in his thesis for the National Institute of Health and Medical Research), 'cancer of the colon originates nearly always from a polypus, that is to say, a small benign growth'. Over the past fifty years, one Frenchman out of ten had polypus growths on the intestinal wall, and when one knows that one polypus out of twenty degenerated into a malignant tumour (cancer), it is somewhat worrying.

Polypi are easily spotted on an X-ray of the intestine after a barium enema. They can then be removed by fibroscopy without surgical intervention, thanks to an instrument perfected for the job called a coloscope. But a slow diagnosis can lead to crisis, which is really regrettable, especially when you think that a healthy diet is enough to prevent this type of cancer.

The main way to prevent cancers of the digestive tube must be the omission of foods which cancerologists agree are carcinogenic. With alcohol responsible for cancer of the oesophagus, and peanuts which contain aflatoxins suspected for provoking cancer of the liver (you should never give peanuts to children), cancerologists put the blame on excessive consumption of fats and animal proteins for cancers of the digestive tract. Animal fats break down during digestion into saturated fatty acids which it is certain are carcinogenic – from now on use vegetable oil or margarine rather than butter – there are some excellent ones available in health food shops and general supermarkets. Animal proteins break down into amino acids, certain of which have metabolisms predisposed to cancer.

The area of research which excites cancerologists working on dietary prevention of cancer of the colon, and this was one of the conclusions that the Lyons colloquy had already reached, compares dietary content of fibre – in particular bran – and the development of cancer of the colon.

Many observations justify this association and this comparison. All epidemiological inquests realizes this rapport between cancer of the colon and poor fibre content in diet. This relation does not have an ethnic or genetic origin, but it rests uniquely on dietary habits. And it is evident that cancer of the colon is directly linked to the rise in the standard of living, to urbanization, and to the complete turnabout in diet which this evolution has brought about. The work of Burkitt and Trowell, amongst others, has shown this to be correct.

In those who eat little fibre and little bran, the faeces move slowly through the colon and stagnate for whole days at a time. This leads cancerologists to make the following observation. Intestinal flora are disturbed by this stagnation of stools in the colon and become predominantly anaerobic: the proportion of aerobic germs noticeably diminishes (up to thirty times, if one compares the colon of an English townsman with that of a Negro countryman). Now, the anaerobic germs which then proliferate have the property to break down bile salts, changing them into carcinogenic substances, in particular into 20-methyl-cholontrine. And this is exactly what happens. The carcinogenic substances from this transformation then stagnate and stay much longer in contact with the intestinal mucosa before being expelled. They must have the opportunity to act for a long time and their harmful effect is intensified. That is why it is so important to introduce bran into our diet to prevent cancer of the colon.

## How to Use Bran
We can clearly distinguish two separate cases:

—People who do not suffer from any particular problem and enjoy relatively good health . . . and want to stay this way.

—People who *are* afflicted with the various problems we have gone into over the last few pages: constipation, gall stones, etc.

**Preventively.** For those people in the first case, an extra helping of bran is not especially justified: that is, bran over and above the amount which they get naturally. But to stay in good health, it is absolutely essential, and in particular if you want to avoid constipation, to eat natural foods rich in bran.

—wholemeal bread (this is the number one food: shut the door once and for all on white bread).
—brown rice and whole cereals in general.
—wholemeal flour, and flakes from whole cereals.

In addition we advise you to introduce into your diet a maximum of fibrous foods, that is to say principally green vegetables: salads, green beans, cabbage, spinach, peas... but also tomatoes, fennel, courgettes, carrots, celery, cucumbers, etc. You will be in excellent health if you get into the habit of using a maximum of herbs in your cooking: garlic, onion, parsley, chervil, basil, tarragon, thyme, sage, rosemary, marjoram, wild thyme, etc.

Finally, only eat meat in tiny quantities once or twice a week, alternating with fish, getting your animal proteins from milk, cheese and eggs instead. Totally eliminate cooked animal fats. Considerably reduce your sugar consumption. And improve your breakfasts mixing fruit and cereals in all sorts of combinations which are made possible by the range of natural healthy foods on offer.

**Curatively**. In the second case, if you suffer from one of the diseases already discussed, in particular - since it is the most common - constipation, you must follow all the points that we have just given in the preceeding paragraph, and add 3 tablespoonsful of bran to your daily diet. You will find organically produced bran in all natural health food shops.

To incorporate your dose of bran easily into your food, without spoiling taste, we advise you to use a product sold in the same shops which is composed of 90 per cent bran and 10 per cent malt. In the form of these granules, it more easily mixes with food than raw bran, and it does not alter the taste.

You can use bran at breakfast, lunch or dinner, mixing your 3 tablespoonsful with different foods:

—in yogurt, or fresh, white soft cheese (cottage cheese).
—in a glass of milk, fruit juice or water.
—in soup.
—in your breakfast cereal, porridge, muesli.
—in your bread dough, adding 30 per cent of bran to the flour.
—in your pastry, when preparing a tart or pudding.

To conclude, a good plan: if you haven't done so already ... REDISCOVER BRAN!!

# 8.

# CEREALS FOR BABY FOOD

### A Baby's Needs

Traditional methods of baby feeding have changed considerably over the last thirty years. Breast-feeding has progressively been abandoned, which is the most important reason for tackling the question of the newly-born's diet. Today we have a far greater knowledge of the digestive capacity of young babies than ever before. Our understanding of the *real needs* of the newly-born baby has become much deeper. The industry responsible for the manufacture of baby food has made enormous progress in the sense of better adapting its products to a baby's needs. Of course, differences of opinion still exist but in the main, a unity of thought has been reached.

In general terms, baby feeding must be approached with liberalism. The new-born baby is perfectly capable of adapting its food supply to its own needs, and these needs vary enormously from one baby to another. There is no need to have a strict rule regarding the quantities you should give: all the examples given in this chapter use average amounts. *You must never force a baby to eat,* the amount he takes can vary from one meal to another, one day to another.

### The Digestive Capacity of a Baby

A thorough digestion of flour by a young baby presents a major difficulty: starch, which is the principal constituent of flour (about 75 per cent) undergoes several transformations during the digestion process, to become a simple and assimilable sugar. Under the action of different digestive enzymes, it changes into dextrine, then maltose and finally glucose. The necessary enzymes for this starch hydrolysis are not secreted in sufficient quantities by a baby until it reaches the age of six months, in particular pancreatic amylase. Practically non-existent at birth, pancreatic amylase appears during the baby's third week, but its activity, which sometimes begins in the third month, normally starts in the sixth month. Thus, logically, you should not give undigested flours before the sixth month; a baby cannot digest them completely and should be given instead roasted, dextrinized and malted flours.

## A Baby's Digestive Intolerance
Giving a baby flour in its diet too soon or in too great a quantity, often results in dyspepsia. This digestive intolerance manifests in abundant, soft, frothy and acidic stools. These abnormal stools are accompanied by a sore bottom, agitation and insomnia. This dyspepsic reaction to flour only occurs if you give your baby too much, which is usually because you decide to increase the manufacturer's recommended amounts, which are in general already quite large enough.

However, there is a more serious digestive intolerance: it is the coeliac problem, which is a manifestation of an intolerance to a gluten amino acid, gliadin.

If your baby suffers from this, you must *never use normal flours – only use gluten-free flour*.

## The Dietetic Benefits of Flour for Infants
The introduction of flour into a baby's diet has several advantages:

—the early variation of his diet means that it is more likely to be truly adapted to the real needs of his body;
—this variation avoids deficiencies which are the result of a too-prolonged diet of pure milk;
—flours facilitate the digestion of milk by modifying its coagulation in the stomach, allowing the digestive enzymes to work more thoroughly;
—flours provide a valuable calorie supplement, without raising the baby's hydrous intake;
—flours encourage a better secretion of pancreatic amylase.

RYE

Thus, after milk, the most vital food in the first weeks of life is flour and you should not hesitate to use it, in milk mixtures from two months, in porridge after three months.

Let's not ignore another factor justifying the early introduction of flour into a baby's diet: towards the age of three months, a baby becomes capable of swallowing without sucking. In a newly-born baby, the passage of food from mouth to stomach is a reflex action and the dissociation of this automatic reaction happens around three months. You should take advantage of this dissociation of sucking and swallowing by beginning to give your baby food with a spoon. The transfer of suction to the spoon usually happens without problem at this age. It becomes more difficult at six months and progressively more difficult thereafter. Spoon-feeding has a great educative benefit, in as much that it allows the baby to recognize changes in food consistency.

## Different Types of Flour for Infants

*Variations*
There are single cereal flours and blended flours:

—cereal flours: wheat, rice, barley, oats, rye, maize. Millet and buckwheat are rarely used for babies.
—blended cereal flours: two or several cereals are mixed together.
—cereal and soya flours: they contribute protein, which is useful for some babies.

—cereal and vegetable flours: one or several vegetables are mixed into the flour, green beans, leaks, parsley, spinach, etc.
—cereal and fruit flours: all combinations are possible, with oranges, apples, bananas, etc.
—flours enriched with cassava or arrow root.

*Types*
There are both simple flours and milky flours. The first are prepared with milk, or later vegetable broth, but you

never prepare the second with milk, or you overfeed which is prejudicial to your baby's stable health. Milky flours should not be used in the first few months.

*Cooking*
Progress in the food industry has made pre-cooked and instant flours more generally available, and these are far easier to prepare than normal flours.

*Predigested Flours*
For the reasons we have given in the second and third parts of this chapter, the manufacture of pre-digested flours has considerably increased over the last few years. There are two sorts:

—dextrinized flours. These are flours which have been roasted. The starch is transformed into dextrine by the roasting which aids a baby's digestion.
—malted flours. We indicate elsewhere the remarkable properties of malt (roasted, germinated barley). The starch is changed into maltose, which is more digestible and more easily assimilated.

**Ways to Use Flour For Babies**
Flours are used in milk mixtures and in porridge:

—Milk Mixes. These pave the way for cutting baby's daily milk feeds between two and three months. They are extremely liquid porridge (see the following table for amounts).
—Porridge. Preparation depends on which flour is used.
—Normal flour. Soak the flour in a little cold water, then add to preheated milk. Bring to the boil slowly, stirring with a wooden spoon, and then gently simmer for 10 minutes. The length of cooking depends on how thick you want your porridge; whether your baby is going to have it in a bottle or with a spoon. This is the best porridge because it has an oily texture.

But you don't always have the time to prepare porridge in this way. It is practical to be able to use precooked varieties sometimes.

—Pre-cooked flour. Proceed as before but limit the simmering time to about 2 minutes.
—Instant flour. Mix the flour directly into the hot milk in the feeding bottle and give it a good shake to mix.

It is important to remember the names of the flour for infants. Sometimes it is called 'cream' or 'flowers'. These types of flour are finer ground than usual, but you use them in the same ways.

## Flour Quantities to Use During the First 6 Months

—2 months: In total for the day, 1 teaspoonful of flour should be mixed with milk.
—2½ months: 2 teaspoonsful maximum for a day, still in a milk mix.
—3 months: Introduction of runny porridge given in the feeding bottle. 3 teaspoonsful for a day.
—4 months: You can start giving your baby 5 feeds, 4 hours apart, and replacing 1 bottle feed with porridge with a spoon. The total quantity of flour for the day should not exceed 12 grams.
—6 months: Introduction of milky flours. About 15 grams – more or less, depending on the baby – in a day.

*Note:* From 3 to 6 months, to thicken the bottle feeds, incorporate cooked vegetable purées (take care to strain the purée before adding to the bottle).

## Flour/Vegetable/Fruit/Mixtures

From three months, to vary your baby's diet and to provide him with mineral salts and natural vitamins, you can prepare clever mixes by adding cooked vegetable purées or stewed fruits, in small quantities to flours.

According to the season, and the fresh vegetables you can get, you can add to the bottle-feed carrot purée, a few cooked, chopped lettuce leaves, or spinach (always cooked, chopped and sieved), or green beans.

One could – or should, even – add to the bottle feeds from time to time several drops of fruit juice (orange, pineapple, grapefruit etc.) or several spoons of apple sauce or mashed poached banana.

## Using Cereal Grains

To naturally correct certain problems encountered with young babies, make use of cereal grain milk mixes which allow you to cut down the bottle feeds.

Boil a tablespoonful of grains in 2 pints (1 litre) of water for an hour. Then push the lot through a sieve. Add to the different bottle feeds during the day.

—for constipation use barley.
—for soft stools use rice.
—for lack of energy use oats.

To make chewing easier, after baby's first teeth are through, don't hesitate to use cereal flakes. Also don't forget that oats encourage milk production: mothers who are breast-feeding should eat oats in preference to other cereals!

# 9.

# THE BY-PRODUCTS OF CEREAL PROCESSING

## Flours

The transformation of cereals into flour takes two operations: *milling* which consists of crushing the grains, then grinding them, traditionally with stone wheels, and *bolting* which separates, by sifting, the almond flour and the husk flour.

Since time began, millers, in their windmills, have ground grain with stone wheels: sandstone, lava, limestone, or granite. The wheels always go in pairs: the incumbent wheel, smaller, immobile, and the running wheel, larger, receiving through a system of gearwheels the transmission of power. By modifying the alignment of the wheels, you can get: either a 'high mixture', rich in wheat flour (fragments of grain) and in bran, and poor in flour; or a 'low mixture' (by bringing the wheels closer together), rich in flour and poor in wheat flour. The main advantage of this traditional method is that it does not overheat the flour and preserves its vital elements, because the wheels turn slowly. Modern industrial milling, by rollers, a totally different process, became generally used at the beginning of this century. The grains pass through a waterfall of rollers, some rough, then smooth ones. The grains are then crushed by fast rollers which move closer together. The flour is severely affected by this process because it becomes too hot through the very fast movement of the rollers. The progress achieved by modern milling methods thus goes against the quality of the flour – in dietetic terms – by sacrificing product to productivity. It is the consumers who are paying for this evolution, and even more so because the milling trade have, for the past thirty years, been concentrating on the manufacture of white flour.

It is in fact more difficult to keep whole flour. It goes off more quickly than white flour because it contains milled germ, which is rich in lipids, as we know. White flour, stripped of its germ and bran, keeps, transports and stocks better, and it lends itself perfectly to all the modern industrial processes. But we have continuously shown in this book what a mistake the consumption of refined cereals can be.

The scale of extraction of flour, still called the bolting scale, is the proportion of whole grain left in the flour. The higher up the scale of extraction the flour is, the more 'whole' it is (rich in bran and the products of milling the protein layer and germ). The lower down the scale, the more white and refined the flour. There are several 'types' of flour:

| Flour Denomination | Scale of Extraction |
| --- | --- |
| Type 45 | 68% grain |
| Type 55 | 74% grain |
| Type 65 | 78% grain |
| Type 80 | 82% grain |
| Type 110 | 85% grain |
| Type 150 | 94% grain |

The flour most frequently used to make the wholemeal bread that you find in health food shops, is type 80 or 110. Of course, it is to be hoped that this flour is the product of wheat cultivated without chemicals. Amongst these flours is Borsa flour which is obtained by a new process which allows – by abrasion – the splitting of the cells of the germ's protein layer – which makes them totally digestible.

It is necessary to take care, before making flour, to clean the grains perfectly. They contain, in fact, a large quantity of foreign bodies from the fields where the cereals grow: wild grains, dust, mud, stones, twigs, insects, etc. It is important to sort out the grains in a sieve and then to brush them thoroughly. These operations are absolutely necessary to ensure sufficiently pure flour.

In addition to breadmaking, which is flour's main market, and (in order of importance) to cake and pastry-making and baby food manufacturing, to cite only two examples, flour is nowadays used more and more for rusks and biscuitmaking.

## Rusks

With a very real growth in the amount of sitting down we do, the consumption of rusks and toast, for roughage, has developed. Rusks, etymologically speaking, are cooked twice, and this double cooking has dietetic advantages: the starch is dextrinized and pre-digested. What is more, rusks, because they are harder and drier than bread, necessitate more thorough chewing. Thus they are more digestible when they leave the mouth and sometimes are more suitable for those who suffer from indigestion when eating bread.

Health food shops offer a wide range of rusks suited to different diets:

—wholemeal rusks, for those suffering from constipation.
—seaweed rusks, recommended during the cure of anorexia.
—rusks without salt, for high blood-pressure sufferers and all those who suffer from oedemas.
—rusks with gluten, which are less rich in glucids and are suitable for diabetics.
—rusks enriched with wheatgerm or vitamins, suitable for many special diets.

These rusks should always be made with organically grown flour: *We can never repeat this enough.*

## Biscuits

A great variety of biscuits exists on the market and there are constantly new additions. This is a clear indication of the consumer attraction of biscuits, but it is a justified attraction because of the benefits presented by this unique food. Biscuits are agreeable to the taste, keep well, are rich in nutritive elements and in calories, all packed into a small volume. They are ideal food for

BUCKWHEAT

children's snacks, for sportsmen, for trips and travelling. In short, a food which symbolizes perfectly our civilization of leisure!

Biscuit composition varies enormously according to type and quality, but all have three basic ingredients: flour, sugar and fat, to which you can add what you like: milk, eggs, chocolate, spices, dried fruit . . . and unfortunately today, more and more artificial colourings, flavourings, emulsifiers, perservatives . . . Thus one should be extremely vigilant in reading labels, to see the ingredients quickly before buying anything, no matter where. Biscuits sold in health food shops are made with organic flour, raw cane sugar, vegetable oil and natural products, without chemical additives. They thus offer an excellent guarantee in comparison with the biscuits one usually finds in general shops, biscuits composed of an amazing number of superfluous 'E' additives which the manufacturers who produce for the specialist shops, on point of honour, do not use! There are three sorts of biscuits, each made with a different sort of dough:

—a stiff dough gives dry biscuits.
—a soft dough gives rich biscuits.
—a wet dough gives wafer biscuits.

The kneading, shaping, baking, finishing, the whole process of biscuit manufacture, becomes more advanced and more perfect each year to satisfy a more and more demanding customer.

Even more than with rusks, the biscuits you can find in the health food stores are suited to all sorts of tastes and diets. To illustrate this variety, we can give a glimpse of the existing range:

Wheat biscuits, with sesame seeds, millet, buckwheat, sour barm without sugar, millet without salt, bran, linseed grains, pollen, with chocolate, soya enriched with raisins, the same without salt and sugar, macaroons with coconut, iced gingercakes, orange-iced gingercakes, dried fruit biscuits, neopolitans, madeleines with butter, shortbread, Normandy gateaux with orange flower honey, savoury snack biscuits, cheese biscuits enriched with wheatgerm . . .

You can also find different breads, with spices, rye and honey, with a slow leaven fermentation and baked at a low temperature. These are the old fashioned recipes of years ago when such baking was a real art – all achieved with natural organically grown products, and certainly without chemical additives.

## Pasta

Quite a large quantity of cereals are consumed in the form of pasta (macaroni, pasta shells, vermicelli, spaghetti and noodles). These pastas are made with a particular type of flour, rich in gluten, originating from a mixture of special wheat varieties: 'hard' wheat. The dough obtained from these wheats has an elasticity and a stickiness so that it can be easily stretched. Some pasta has eggs added, which makes it an even more nourishing food, but it is necessary for those who suffer from diabetes or obesity to eat it with caution: those who eat large amounts of pasta are often very fat!

You can find wholemeal pasta in health food shops and also pasta enriched with gluten, less rich in carbohydrates, which suit diabetics.

## Breakfast Cereals

Those of you who have read this far and have still not changed your diet for the better, may like to start by rethinking your breakfasts: it can be at breakfast, at the beginning of each day, that you can exercise your will-power to change your eating habits. Breakfast should be the first step towards a correct diet. You should recognize that, in general, this meal is totally ignored by the majority of people, and what a mistake! Think of those milky coffees and white toast and jam, eaten in a rush . . .

Now, for those of you who today have decided to do something about your diet, it is no longer a question of a breakfast such as above. And since the fact that you have bought this book shows a step towards an interest in cereals, cereals should henceforward have a privileged place at breakfast. You should understand that cereals lend themselves remarkably well to the preparation of well-balanced, nutritious and energy-giving breakfasts, which are most important, especially for children. They can also be prepared in a great many ways, as we will show in the few examples that we give at the end of the chapter. It is important for those who are new to this type of diet that their food should taste good, so attention should be paid to what you put with your cereals.

You can eat cereals in several forms at breakfast:

—in flakes used for porridge (we give the recipe further on) or muesli.
—puffed cereals.
—flours.
—wheatgerm, which you can have mixed with what you prepare and which is especially good for children.
—in the form of sprouted and toasted cereals: malt or different coffee substitutes based on barley and rye . . . which you can find in health food shops.

Here are a few recipes to cheer up your cereal breakfasts:

## (a) Porridge

*Basic recipe:* slowly stir 5 oz/150g (1¼ cupsful) of rolled oats (large or small) in 1¾ pints/1 litre (4½ cupsful) of boiling salted water. Cook for 5 minutes over a low heat stirring constantly with a wooden spoon – and then remove from the heat. You will then have quite a thick porridge.

Porridge is mainly eaten at breakfast time, and especially in winter – but you could, of course, eat it at night! The basic recipe can easily be varied to relieve the boredom!

—with honey.
—with cinnamon or vanilla.
—with a zest of lemon or orange, if you are sure that the fruit has not been treated to preserve it (always buy your fruit in health food shops).

—with coconut or ground almonds.
—with dried raisins (sultanas or currants) and little pieces of dried fruit (such as angelica).
—with grated apple.
—with red fruit jelly (redcurrant, strawberry, bilberry, blackcurrant).
—with fresh fruit purée (banana or peach).

## (b) The Natural Breakfast

| Imperial/Metric | American |
| --- | --- |
| 7 oz (200g) wholemeal flour (freshly ground) | 1¾ cupsful wholewheat flour (freshly ground) |
| 2 oz (50g) raisins | ⅓ cupful raisins |
| 1 pint (570ml) water | 2½ cupsful water |
| 1 tablespoonful wheatgerm | 1 tablespoonful of wheatgerm |
| 2 oz (50g) ground almonds | ½ cupful ground almonds |
| 2 teaspoonsful almond purée | 2 teaspoonsful of almond paste |
| 4 apples | 4 apples |

1. The previous evening, put the flour and the raisins to soak in water. (If you have a cereal mill, grind the wheat at the last moment.) Leave to stand overnight.

2. In the morning add the wheatgerm, the ground almonds and purée (paste).

3. Mix well before adding – at the last minute – the grated apple.

*Note:* The ideal drink to accompany the naturist breakfast is malt. We would also advise that you chew this meal very carefully!

## (c) Puffed Cereals

You will find in all health food shops, natural puffed cereals. The most common are wheat, barley and rice.

These puffed cereals give you the opportunity to vary your breakfasts (by alternating with porridge and muesli), all using an appreciable quantity of cereals. They can be served in a variety of ways:

—in yogurt or fresh soft cheese (such as cottage cheese).
—in milk and honey.
—mixed with fresh fruit purée or fruit compote.

## (d) Muesli

*Basic recipe:* the previous evening soak 4 tablespoonsful of cereals (1 of wheat, 1 of barley, and 2 of oats) in a bowl of water. That's all – childishly easy and yet this elementary recipe forms the basis of what is perhaps the most agreeable and varied way of eating cereals. Innumerable mueslis can be made from this basic recipe of cereals which you simply have to soak overnight.

Here are a few suggestions:

*Muesli With Dried Fruit:* the previous evening soak separately several dried fruits: prunes, apricots, sultanas and figs. In the morning mix the chopped fruits with cereals, adding two or three dates (which do not need to be soaked before hand), diluting the lot with a little milk (or almond milk or soya milk) and honey.

*Muesli With Bran:* we recommend this, naturally, to those who suffer from constipation. The night before, leave cereal flakes, a few prunes and a tablespoonful of linseed grains to soak in separate bowls. In the morning, mix everything together, adding 2 tablespoonsful of bran and 2 tablespoonsful of rhubarb jam, diluting the lot with barley water.

*Muesli With Apple:* in the morning mix cereal flakes with grated apple (only grate at the last moment to keep it from browning) and 1 teaspoonful of lemon juice plus a few drops of vanilla.

*Muesli With Almonds:* mix cereal flakes with freshly grated almonds, walnuts, cashews, hazelnuts, pistachios . . . diluting with almond milk.

*Muesli With Red Fruit:* crush a selection of red fruits (according to season) and mix the juice and pulp with cereal flakes. Mulberries, redcurrants, blackberries, strawberries, raspberries, bilberries. Serve with grapefruit juice.

# 10.

# CEREAL RECIPES

To familiarize you with the various cereals, we have given you several basic recipes. The proportions of the ingredients are given for four people in all the recipes which follow. Here are several hints on the products to use:

—In every case, whether you use cereal grains, flours or flakes, always choose whole cereals or products derived from whole cereals. Refined products, such as white rice and white flour, should absolutely be avoided.

—When you are buying cereals, make sure that they are always organically grown: this is most important for your health because whole cereals which have not been grown organically contain all the residues of chemical treatments.

—Vegetables which accompany certain dishes should also be organically grown wherever possible.

—Fruits used in the desserts must be as natural as possible: fruits which grow in this country (apples, pears, plums, etc.) should be organically grown; imported citrus fruits (lemons, oranges, grapefruits, etc) without preservatives; if possible dried fruits without preservatives (this is possible for prunes - one should choose Agen prunes rather than Californian for this reason - but unfortunately impossible for apricots, for example, which are always sulphured), and preserved fruits without colouring.

—Always buy sea salt, not refined.

—Use raw cane sugar, not refined white, and when you can, replace sugar with honey.

—The honey you use should be pure, natural and not heat treated.

—Use pure olive oil from a first cold pressing, not refined, and it should have an acidity lower than 0.5.
—Vegetable oil should be pure palm oil, not hydrogenized.
—Buy untreated spices, or better still grow and gather your own. What can be more satisfying than spending a Sunday or holiday looking for thyme, rosemary, savory...

*Note:* You will find all these wholesome products in your local health food shop.

## Soups and Stews

### Roasted Cereal Flake Stew

| Imperial/Metric | American |
|---|---|
| 4 tablespoonsful of cereal flakes (barley, oats, rice or wheat) | 4 tablespoonsful of cereal flakes (barley, oats, rice or wheat) |
| Olive oil | Olive oil |
| 4 pinches of sea salt | 4 pinches of sea salt |
| 1¾ pints (1 litre) of water | 4½ cupsful of water |
| 2 egg yolks | 2 egg yolks |
| Parsley | Parsley |
| Chervil | Chervil |

1. Sauté the cereal flakes in a frying pan (skillet) until lightly browned.

2. Carefully dilute with the cold salted water.

3. Slowly bring to the boil and remove from the heat as soon as boiling point is reached.

4. Thicken with the beaten egg yolks.

5. Serve sprinkled with parsley and chopped chervil.

## Vegetable and Cereal Flake Soup

| Imperial/Metric | American |
| --- | --- |
| 2 carrots | 2 carrots |
| 1 leek | 1 leek |
| 1 stick of celery | 1 stalk of celery |
| 2 turnips | 2 turnips |
| Olive oil | Olive oil |
| 1¾ pints (1 litre) of water | 4½ cupsful of water |
| 4 pinches of sea salt | 4 pinches of sea salt |
| 4 tablespoonsful of cereal flakes (barley, oats, rice or wheat) | 4 tablespoonsful of cereal flakes (barley, oats, rice or wheat) |
| 2 tablespoonsful of fresh cream | 2 tablespoonsful of fresh cream |
| Parsley | Parsley |

1. Finely chop the vegetables and sauté for several minutes in the oil.

2. Add the water, salt, cereal flakes and gently simmer for 30 minutes.

3. Add the cream and chopped parsley just before serving.

## Provence Cereal Soup

| Imperial/Metric | American |
| --- | --- |
| 1 lb (½ kilo) tomatoes | 1 pound tomatoes |
| 2 onions | 2 onions |
| Olive oil | Olive oil |
| 2 cloves of garlic | 2 cloves of garlic |
| Basil | Basil |
| Wild thyme | Wild thyme |
| Marjoram | Marjoram |
| 4 pinches of sea salt | 4 pinches of sea salt |
| 1¾ pints (1 litre) of water | 4½ cupsful of water |
| 4 tablespoonsful of cereal flakes (barley, wheat, rice or oats) | 4 tablespoonsful of cereal flakes (barley, wheat, rice or oats) |

1. Sauté the roughly chopped tomatoes and finely chopped onion with the garlic and herbs.

2. Add the salt and water and boil for 15 minutes.

3. Add the cereal flakes and leave to swell for several minutes before serving.

4. Serve with a nob of butter.

## Cream of Vegetable and Cereal Soup

| Imperial/Metric | American |
|---|---|
| 2 carrots | 2 carrots |
| 1 leek | 1 leak |
| 1 stick of celery | 1 stalk of celery |
| 2 turnips | 2 turnips |
| 2 onions | 2 onions |
| Olive oil | Olive oil |
| 2 pints (1¼ litres) of water | 5 cupsful of water |
| 4 pinches of sea salt | 4 pinches of sea salt |
| 4 tablespoonsful of creamed cereals (barley, oats, rice or wheat) | 4 tablespoonsful of creamed cereals (barley, oats, rice or wheat) |
| 1 nut of butter | 1 nut of butter |
| Parsley | Parsley |
| Chervil | Chervil |

1. Sauté the chopped vegetables in the oil.

2. Then add 1¾ pints/1 litre (4½ cupsful) of the water and the salt.

3. Cook for 20 minutes.

4. Stir well and add the creamed cereal which has been diluted with the remaining cold water and brought to the boil.

5. Cook for 3 minutes stirring continuously.

6. Serve with the butter and herbs.

## Spinach and Cereal Flake Stew

| Imperial/Metric | American |
| --- | --- |
| 2 onions | 2 onions |
| Olive oil | Olive oil |
| 1¾ pints (1 litre) of water | 4½ cupsful of water |
| 4 pinches of sea salt | 4 pinches of sea salt |
| 1 pinch of thyme | 1 pinch of thyme |
| 1 bay leaf | 1 bay leaf |
| 1 pinch of rosemary | 1 pinch of rosemary |
| 4 tablespoonsful of cereal flakes | 4 tablespoonsful of cereal flakes |
| 10 oz (300g) spinach | 10 ounces spinach |
| Nutmeg | Nutmeg |

1. Sauté the onions until translucent.

2. Add the water, salt, herbs, cereal flakes and spinach.

3. Gently cook for 15 minutes.

4. Stir well and serve with a sprinkling of grated nutmeg.

## Leek and Cereal Flake Stew

| Imperial/Metric | American |
| --- | --- |
| 4 leeks | 4 leeks |
| Olive oil | Olive oil |
| 1¾ pints (1 litre) of water | 4½ cupsful of water |
| 4 pinches of sea salt | 4 pinches of sea salt |
| 4 tablespoonsful of cereal flakes (barley, oats, rice or wheat) | 4 tablespoonsful of cereal flakes (barley, oats, rice or wheat) |
| 1 pinch of tarragon | 1 pinch of tarragon |

1. Slice the leeks finely and gently sauté for a few minutes.

2. Add the water, salt, cereal flakes and gently simmer for 30 minutes.

3. Stir well and serve with a little finely chopped tarragon.

## Cress Soup

| Imperial/Metric | American |
|---|---|
| 2 onions | 2 onions |
| Olive oil | Olive oil |
| 2 pints (1¼ litres) of water | 5 cupsful of water |
| 4 pinches of sea salt | 4 pinches of sea salt |
| 4 tablespoonsful of creamed cereals | 4 tablespoonsful of creamed cereals |
| 5 oz (150g) cress | 5 ounces cress |
| 1 nut of butter | 1 nut of butter |

1. Sauté the chopped onion.

2. Add 1¾ pints/1 litre (4½ cupsful) of water, salt and creamed cereal which has already been diluted with the remaining water and brought to the boil.

3. Simmer for 3 minutes stirring well.

4. Add the cress and mix thoroughly.

5. Serve with the butter.

# Cereal Burgers

## Cereal Flake Burgers

| Imperial/Metric | American |
|---|---|
| 5 oz (150g) cereal flakes (barley, wheat, oats or rice) | 3 cupsful cereal flakes (barley, wheat, oats or rice) |
| Sea Salt | Sea salt |
| 1 pint (½ litre) of water | 2½ cupsful of water |
| 1 clove of garlic | 1 clove of garlic |
| 1 onion | 1 onion |
| Mixed herbs | Mixed herbs |
| Olive oil | Olive oil |
| 2 eggs | 2 eggs |
| 3 tablespoonsful of wholemeal flour | 3 tablespoonsful of wholewheat flour |

1. Soak the cereal flakes for half an hour in the boiling salted water.

2. During this time sauté the garlic, onion and herbs in the oil.

3. Mix all the ingredients together adding the two whole eggs.

4. Scatter the flour on a plate and take a ball of the mixture.

5. Flatten it between the palms of your hands and dip the burger (about 1 inch/2 cm deep) in the flour.

6. Fry in the hot oil and turn after 5 minutes frying the other side for another 5 minutes over a slightly lower heat.

This is the basic recipe. We personally prefer a fairly thin burger which can be fried in a small amount of oil, to the croquette or steaklet which are thicker and which must be deep fried.

*Note:* The cereal flake burgers can be served on their own, with a green salad, with all sorts of steamed vegetables, with a tomato sauce, with grated gruyère cheese, a nob of butter or melted cheese.

# Brown Rice Burgers

| Imperial/Metric | American |
|---|---|
| 1 cupful of cooked brown rice | 4/5 cupsful of cooked brown rice |
| 2 eggs | 2 eggs |
| 2oz (50g) grated gruyère | ½ cupful grated gruyère |
| 4 tablespoonsful of wholemeal flour | 4 tablespoonsful of wholewheat flour |
| Breadcrumbs | Breadcrumbs |
| Olive oil | Olive oil |

1. Mix the rice, eggs, cheese and flour thoroughly.

2. Make burgers as in the previous recipe and cover with breadcrumbs.

3. Fry in the oil for 5 minutes on each side.

# Wheat and Tomato Burgers

| Imperial/Metric | American |
|---|---|
| 6 tomatoes | 6 tomatoes |
| 3 spring onions | 3 scallions |
| 12 tablespoonsful of wheat flakes | 12 tablespoonsful of wheat flakes |
| Basil | Basil |
| 2 egg yolks | 2 egg yolks |
| Sea salt | Sea salt |
| Breadcrumbs | Breadcrumbs |
| Olive oil | Olive oil |

1. Sauté the finely chopped tomatoes and spring onions.

2. Remove from the heat and add the wheat flakes, the chopped basil, egg yolks and salt.

3. Leave to swell for a few minutes.

4. Make the burgers with the mixture and dip them in the breadcrumbs.

5. Fry them in the oil for 5 minutes each side.

## Wheat and Mushroom Burgers

| Imperial/Metric | American |
|---|---|
| 5 oz (150g) wheat flakes | 3 cupsful of wheat flakes |
| 1 pint (½ litre) of water | 2½ cupsful of water |
| Sea salt | Sea salt |
| 9 oz (250g) mushrooms | 4 cupsful mushrooms |
| 2 oz (50g) black olives | ½ cupful black olives |
| 1 clove of garlic | 1 clove of garlic |
| 1 onion | 1 onion |
| Olive oil | Olive oil |
| 3 tablespoonsful of wholemeal flour | 3 tablespoonsful of wholewheat flour |

1. Soak the wheat flakes in the boiling salted water for half an hour.

2. During this time sauté the finely chopped mushrooms, olives, garlic and onion.

3. Mix in the soaked wheat flakes.

4. Make burgers, dip them in the flour and fry them in hot oil for 5 minutes each side.

## Vegetable Pâtés

### Vegetable Pâté with Wholemeal Bread

| Imperial/Metric | American |
|---|---|
| 9 oz (250g) wholemeal bread | 9 ounces wholewheat flour |
| Water | Water |
| 4 spring onions | 4 scallions |
| 1 clove of garlic | 1 clove of garlic |
| Parsley | Parsley |
| Chervil | Chervil |
| Tarragon | Tarragon |
| Basil | Basil |
| Olive oil | Olive oil |
| 2 eggs | 2 eggs |
| Sea salt | Sea salt |

1. Soak the bread – cut into small cubes – in water.

2. During this time, sauté the garlic and finely chopped herbs.

3. Mix well with the bread and the eggs until it all sticks together. Add the salt.

4. Turn the mixture into a small greased loaf tin and bake at 425°F/220°C (Gas Mark 7) for 45 minutes.


## Spring Vegetable Pâté

| Imperial/Metric | American |
|---|---|
| 1 handful of nettle leaves | 1 handful of nettle leaves |
| 1 handful of radish tops | 1 handful of radish tops |
| Several melissa leaves | Several melissa leaves |
| 1 handful cress | 1 handful cress |
| Water | Water |
| Sea salt | Sea salt |
| 9oz (250g) wholemeal bread | 9 ounces wholewheat bread |
| 2 cloves of garlic | 2 cloves of garlic |
| 2 sticks of celery | 2 stalks of celery |
| Olive oil | Olive oil |
| 2 eggs | 2 eggs |
| Rosemary | Rosemary |
| Sage | Sage |
| Savory | Savory |

1. Plunge all the greens into 1 pint/½ litre (2½ cupsful) of boiling salted water and stir well.

2. Leave the bread to soak in this herby soup, but do not cook.

3. During this time, sauté the garlic and celery in the olive oil.

4. Mix everything together adding the eggs and herbs at the last moment.

5. Turn the mixture into a small loaf tin and bake in a hot oven for 45 minutes.

## Main Dishes

### Brown Rice

*Basic recipe:* There are a number of ways of cooking brown rice, but we have chosen the one which always turns out perfectly:

| Imperial/Metric | American |
|---|---|
| 9 oz (250g) brown rice | 1¼ cupsful brown rice |
| 1¾ pints (1 litre) of water | 2½ cupsful of water |
| 1 teaspoonful of sea salt | 1 teaspoonful of sea salt |

1. Rinse the rice thoroughly in cold water stirring well.

2. Tip the rice into the boiling salted water.

3. Leave to simmer for half an hour with the lid on.

4. Turn off the heat and leave to swell for half an hour without lifting the lid.

If you have never cooked brown rice, it will only be after two or three attempts that you will get it just right. Because there are so many varieties of rice, each type takes a slightly different method of cooking. If you find a rice which suits you (at any rate, always buy brown rice from health food shops who stock organically grown products), don't change your brand or variety: then the result will always be the same. With brown rice, one can always add all sorts of ingredients to make a variety of dishes, paella, risotto, Indian rice, Chinese rice, Spanish rice . . . There are numerous possibilities provided you have an imagination and know what goes together.

In general it is a good idea to always bear the following proportions in mind in your recipes: 75% rice and 25% of other ingredients which could include:

—carrots, courgettes, petits pois, celery.
—seaweeds.
—mushrooms and olives.
—tomatoes and complementary herbs, basil, rosemary, thyme.

In each case the addition of spices can change the effect: saffron, curry, nutmeg, paprika, etc.

## Millet Ratatouille

| Imperial/Metric | American |
|---|---|
| 7 oz (200g) millet | 1 cupful millet |
| Sea salt | Sea salt |
| 1 pint (½ litre) of water | 2½ cupsful of water |
| 2 onions | 2 onions |
| 2 aubergines | 2 eggplants |
| 4 tomatoes | 4 tomatoes |
| 2 leeks | 2 leeks |
| 4 courgettes | 4 zucchini |
| 4 cloves of garlic | 4 cloves of garlic |
| 2 sticks of celery | 2 stalks of celery |
| Olive oil | Olive oil |

1. Cook the millet very gently in the boiling salted water for 30 minutes.

2. During this time gently sauté all the chopped vegetables in the oil.

3. As soon as the millet is cooked, mix it with the vegetables.

4. Turn off the heat and leave to swell for 5 minutes before serving.

## Saffron Rice

| Imperial/Metric | American |
| --- | --- |
| 2 onions | 2 onions |
| Olive oil | Olive oil |
| 7 oz (200g) brown rice | 1 cupful brown rice |
| Pinch of saffron (in powder or threads) | Pinch of saffron (in powder or threads) |
| 1¼ pints (¾ litres) of water | 3 cupsful of water |
| Sea salt | Sea salt |
| Basil | Basil |

1. Sauté the finely chopped onion in two tablespoonsful of olive oil.

2. Add the rice and saffron and leave to colour for 5 minutes, stirring continuously.

3. Add the cold water and salt and leave to cook at 225°F/110°C (Gas Mark ¼) for an hour.

4. Serve with chopped basil leaves.


## Tomatoes Stuffed With Oat Flakes

| Imperial/Metric | American |
| --- | --- |
| 4 oz (100g) oat flakes | 2½ cupsful oat flakes |
| ½ pint (¼ litre) of water | 1⅓ cupsful of water |
| 9 oz (250g) mushrooms | 4 cupsful mushrooms |
| 4 spring onions | 4 scallions |
| Tarragon | Tarragon |
| Basil | Basil |
| Olive oil | Olive oil |
| 8 tomatoes | 8 tomatoes |
| Sea salt | Basil |
| Parsley | Parsley |

1. Leave the oats to soak in the water for half an hour.

2. During this time, sauté the chopped mushrooms, spring onions and herbs.

3. Cut the tops off the tomatoes and scoop out the pulp.

4. Add the pulp to the oats and mushroom mixture and fill the tomato shells. Add the salt.

5. Place the tomatoes in an oven-proof dish with a drop of water in the bottom and bake in a medium oven for 20 minutes.

6. Serve sprinkled with parsley.

## Polenta

*Basic recipe:*

| Imperial/Metric | American |
|---|---|
| 9 oz (250g) polenta (maize semolina) | 9 ounces polenta (maize semolina) |
| 1¾ pints (1 litre) of water | 4½ cupsful of water |
| Sea salt | Sea salt |
| Olive oil | Olive oil |

1. Gradually pour the semolina into the boiling salted water. Shake vigorously until the water comes back to the boil.

2. Lower the heat and leave to cook for 20 minutes stirring frequently so that lumps do not form. (Be careful that the semolina does not stick to the bottom and sides of the pan.)

3. Turn the polenta into a large greased dish and leave to go cold by which time it should be firm.

4. Cut into cubes and sauté quickly in the oil until golden.

Polenta can be served with:

—steamed vegetables.
—grated gruyère or better still, parmesan.
—sautéed vegetables, mushrooms and onions.
—an omelette aux fines herbs or with sorrel.

## Kasha With Mushrooms

| Imperial/Metric | American |
| --- | --- |
| 9 oz (250g) mushrooms | 4 cupsful mushrooms |
| 4 onions | 4 onions |
| Olive oil | Olive oil |
| 9 oz (250g) kasha (grains of buckwheat) | 9 ounces kasha (grains of buckwheat) |
| Mixed herbs | Mixed herbs |
| Sea salt | Sea salt |
| 1 pint (½ litre) of water | 2½ cupsful of water |
| Parsley | Parsley |
| Chervil | Chervil |

1. Sauté the sliced mushrooms and finely chopped onions.

2. Then add the buckwheat and herbs, mixing well.

3. Add the salted cold water and quickly bring to the boil stirring continuously.

4. Cover and reduce heat, simmer for 15 minutes.

5. Serve with parsley and chopped chervil.


## Buckwheat Crêpes

| Imperial/Metric | American |
| --- | --- |
| 1 lb (500g) buckwheat flour | 4 cupsful buckwheat flour |
| 4 eggs | 4 eggs |
| 4 pinches of sea salt | 4 pinches of sea salt |
| 1 tablespoonful of maize oil | 1 tablespoonful of maize oil |
| 3 pints (1½ litres) of liquid (water, or half water and half milk, or beer) | 7½ cupsful of liquid (water, or half water and half milk, or beer) |

*Preparation of batter:*

1. Mix all the ingredients in an earthenware bowl, working the batter very carefully to prevent the formation of lumps.

2. Add the chosen liquid gradually.

3. Leave the batter to rest for half an hour to an hour

(except when using beer when you do not need to wait).

*Cooking the Crêpes:*

1. Use an iron frying pan (skillet), and heat the pan well.

2. Add the maize oil and heat.

3. When the oil is very hot, pour in a small spoon of batter quickly and tilt the pan so that the base of the pan is covered.

4. Cook the crêpe over a strong heat and turn carefully with a spatula.

This is the basic recipe, but these buckwheat crêpes can be added to in many ways:

—serve them with steamed vegetables.
—stuff them with green vegetables purée (spinach, sorrel).
—stuff them with mushrooms or pitted olives.
—cover them with béchamel sauce, grated gruyère, and bake them.
—cover them with a tomato sauce, parsley, chervil.

If you wish to serve crêpes for dessert, add several drops of orange flower water to the batter, or vanilla or cinnamon. They can then be served in a number of ways:

—rolled in acacia honey.
—rolled with jam or plum purée.
—rolled with almonds and ground nuts.
—rolled with raisins, sultanas (previously soaked in tea) and thin slices of banana.

## Fresh Pasta Made With Wholemeal Flour

*Preparation of Pasta Noodles*

| Imperial/Metric | American |
|---|---|
| 11 oz (300g) wholemeal flour | 2¾ cupsful wholemeal flour |
| 3 eggs | 3 eggs |
| A little water | A little water |

1. Put a little flour on the table. Make a well and break the eggs into it.

2. Work the flour with your fingers and add just enough water to make a firm and elastic dough. When the dough is quite firm, knead it carefully with your palms (the work surface and hands must constantly be covered in flour so that the dough doesn't stick).

3. After 15 minutes of kneading, make a ball with the dough and leave it to rest for 1 hour.

4. After this time lightly rework the dough and make little balls the size of an egg. Stretch out the dough balls with a rolling pin in sheets as thin as possible (about 2mm).

5. Cover a couple of kitchen chairs with clean cloths and lay these thin sheets of dough over them.

6. Leave them to dry for 30 minutes and then cut into noodles with a knife.

*Cooking Fresh Pasta:* Plunge the pasta into 5 pints/2½ litres (12 cupsful) of salted boiling water and cook for 10 minutes.

You can serve this pasta in three main ways:

—a l'anglaise: simply drained, rinsed in boiling water and serve with a nob of butter.
—au gratin: drained, put on a buttered dish with a béchamel sauce, grated cheese and breadcrumbs on top, and baked in a hot oven for 15 minutes.
—a l'italienne: served with a tomato sauce and herbs.

# Cannelloni

| Imperial/Metric | American |
| --- | --- |
| 7 oz (200g) cannelloni | 7 ounces cannelloni |
| 1 onion | 1 onion |
| 1 handful of sorrel | 1 handful of sorrel |
| 1 handful of cress | 1 handful of cress |
| 1 handful of spinach | 1 handful of spinach |
| 4 oz (100g) wholemeal breadcrumbs | 2 cupsful wholewheat breadcrumbs |
| ½ glass of milk | ½ glass of milk |
| Breadcrumbs | Breadcrumbs |
| Grated gruyère | Grated gruyère |
| Oil | Oil |
| Sea salt | Sea |

1. To make cannelloni, cut the sheets of dough into rectangles 3½ x 2 inches (9 x 5 cm).

2. Poach the pieces of pasta in boiling salted water for 10 minutes and lay them out on a cloth to drain.

3. Whilst the cannelloni is cooking, chop the onion roughly, the sorrel, the cress and the spinach and mix together with the breadcrumbs which have been soaked in the milk.

4. Place a tablespoonful of this stuffing on each cannelloni and roll up.

5. Put the cannelloni in a greased baking dish, cover with grated cheese and breadcrumbs and grill for 10 minutes under a strong heat.

# Cereal Quenelles

| Imperial/Metric | American |
|---|---|
| 1½ oz (40g) vegetable oil | 1½ ounces vegetable oil |
| 6 tablespoonsful of wholemeal flour | 6 tablespoonsful of wholewheat flour |
| 2 eggs | 2 eggs |
| Sea salt | Sea salt |
| Nutmeg | Nutmeg |
| 2 tablespoonsful of yeast extract | 2 tablespoonsful of yeast extract |

1. Heat the oil gently and then gradually add the flour mixing well, and then the eggs.

2. Stir for 3 minutes, then remove from the heat and add the salt, nutmeg and yeast extract.

4. Mix well and make into little balls.

*Cooking:* Poach the quenelles in boiling water: tip them into the water and leave them for about 15 minutes, using a draining spoon to lift them out when they rise to the surface.

They can be served with:

—a béchamel sauce.
—tomato sauce.
—mushroom sauce.


# Pizza

*Preparation of the dough:*

1. Take 9 oz/250g of wholemeal flour (2½ cupsful wholewheat flour).

2. Make a well in the middle and pour in 4 oz/100g of vegetable fat and 1 teaspoonful of sea salt.

3. Work these ingredients little by little, gradually mixing in the flour and a glass of water.

4. Knead the dough and leave it to rest for 1 hour.

5. Roll out the dough and place on an oiled and floured tray.

*Preparation of the topping:*

1. Sauté 2 chopped onions and 4 tomatoes in a little oil.

2. Place on the dough and cover with more slices of tomato, grated gruyère or parmesan, olives and a large pinch of marjoram.

*Cooking:* 20 minutes at 425°F/220°C (Gas Mark 7). Before serving sprinkle the surface of the pizza with olive oil which has been mixed with bay leaves, sprigs of thyme, rosemary and a little red pepper.

## Desserts

### Oat Flake Pudding

| Imperial/Metric | American |
| --- | --- |
| 1 pint (½ litre) of milk | 2½ cupsful of milk |
| 1 tablespoonful of oat flakes | 1 tablespoonful of oat flakes |
| 1 egg | 1 egg |
| 1 teaspoonful of acacia honey | 1 teaspoonful of acacia honey |
| 1 pinch of cinnamon | 1 pinch of cinnamon |
| 2 teaspoonsful of raisins and sultanas | 2 teaspoonsful raisins and golden seedless raisins |
| 1 small piece of angelica | 1 small piece of angelica |

1. Heat the milk and gradually add the oat flakes, stirring constantly.

2. Simmer for 5 minutes, stirring all the time over a low heat.

3. Remove from the heat and add the beaten egg, honey, cinnamon and fruit.

4. Turn out into a well greased pie dish placing the angelica on top.

5. Bake for 20 minutes at 350°F/180°C (Gas Mark 4).

## Rice Gateau

| Imperial/Metric | American |
|---|---|
| 4 oz (100g) rice flakes | 2½ cupsful rice flakes |
| 1 pint (½ litre) of milk | 2½ cupsful of milk |
| 1 tablespoonful of honey | 1 tablespoonful of honey |
| 2 eggs | 2 eggs |
| 1 vanilla pod | 1 vanilla pod |

1. Cook the rice very gently in the milk and honey stirring well for 5 minutes.

2. Remove from the heat and add the beaten eggs and vanilla.

3. Turn into a greased mould, place in a double saucepan and bake at 425°F/220°C (Gas Mark 7) for 30 minutes.

## Citrus Fruit Cereal

| Imperial/Metric | American |
|---|---|
| 4 tablespoonsful of wheat flakes | 4 tablespoonsful of wheat flakes |
| 1 pint (½ litre) of orange juice | 2½ cupsful of orange juice |
| 1 tablespoonful of orange blossom honey | 1 tablespoonful of orange blossom honey |
| 2 grapefruits | 2 grapefruits |
| Grated rind of 2 lemons | Grated rind of 2 lemons |

1. Soak the wheat flakes in the orange juice for half an hour.

2. Add the honey and the quartered grapefruits.

3. Serve sprinkled with the lemon rind.

## Petit Fours With Cereal Flakes

| Imperial/Metric | American |
|---|---|
| 7 oz (200g) cereal flakes | 1¼ cupsful cereal flakes |
| 3 tablespoonsful of honey | 3 tablespoonsful of honey |
| 2 tablespoonsful of cocoa powder | 2 tablespoonsful of cocoa powder |
| 1 pinch of cinnamon | 1 pinch of cinnamon |
| Water | Water |

1. Sauté the cereal flakes in a frying pan (skillet) until they begin to brown.

2. Mix all the ingredients with the water until you have a thick paste.

3. Make small balls the size of nuts and leave to dry before storing in a glass jar.

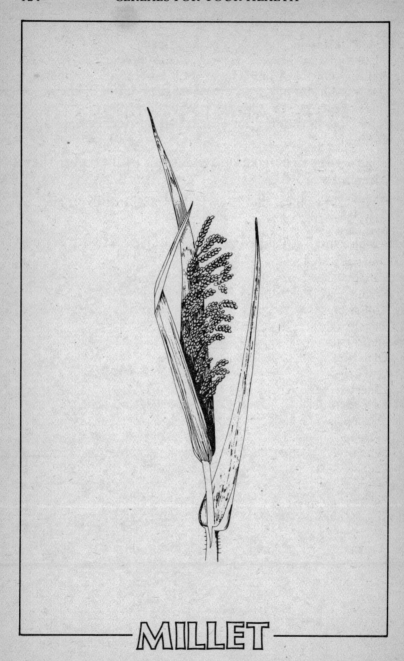

MILLET

# THERAPEUTIC INDEX

| Ailments | Remedies |
|---|---|
| Anaemia | Wheatgerm, Millet |
| Arteriosclerosis | Rye |
| Asthenia | Millet |
| Bladder Inflammations | Barley |
| Breastfeeding | Oats, Buckwheat, Wheatgerm |
| Calcium Deficiency | Wheatgerm |
| Cardiovascular problems | Rye |
| Colostomy | Wheat and Maizegerm oils |
| Constipation | Bran |
| Depression | Millet |
| Diabetes | Oats |
| Diarrhoea | Rice Water |
| Digestive problems | Barley |
| Diverticulitis | Bran |
| Dyspepsia | Malted barley |
| Fatigue | Wheatgerm, raw Wheat |
| Frigidity | Oats, Wheatgerm |
| Gall stones | Bran |
| Growth | Oats, Wheatgerm |
| Haemorrhoids | Bran |
| High Blood-pressure | Rye, Rice (in this case at the exclusion of all other food) |
| Impotence | Wheatgerm, Oats |
| Intestinal Parasites | Raw Rice |
| Lymphatism | Wheatgerm |
| Memory loss | Millet |
| Mineral deficiency | Wheatgerm |
| Nephritis | Rice (to the exclusion of all other food: the Kempner diet) |

| Ailments | Remedies |
| --- | --- |
| Neuritis | Wheatgerm |
| Obesity | Bran |
| Oedemas | Rice |
| Polypi | Bran |
| Pregnancy | Wheatgerm |
| Old Age | Wheatgerm |
| Rickets | Wheatgerm |
| Scurvy | Wheatgerm |
| Skin Inflammations | Rice flour (poultices), Bran (baths) |
| Slimming | Rice |
| Sterility | Oats |
| Thyroid deficiency | Oats |
| Tuberculosis | Wheatgerm |
| Uraemia | Oats |

*Preventive Measures:* for cancer in general: wheatgerm because of its rich magnesium content. For cancer of the colon: whole cereals and bran.

# INDEX